# Advancements in AI and IoT for Chip Manufacturing and Defect Prevention

**Published 2024 by River Publishers**

River Publishers

Alsbjergvej 10, 9260 Gistrup, Denmark

www.riverpublishers.com

**Distributed exclusively by Routledge**

605 Third Avenue, New York, NY 10017, USA

4 Park Square, Milton Park, Abingdon, Oxon OX14 4RN

*Advancements in AI and IoT for Chip Manufacturing and Defect Prevention* / by Rupal Jain.

Routledge is an imprint of the Taylor & Francis Group, an informa business

ISBN 978-87-7004-681-7 (paperback)

ISBN 978-87-7004-683-1 (online)

ISBN 978-87-7004-682-4 (ebook master)

A Publication in the River Publishers Series in Rapids

While every effort is made to provide dependable information, the publisher, authors, and editors cannot be held responsible for any errors or omissions.

# Advancements in AI and IoT for Chip Manufacturing and Defect Prevention

Rupal Jain

MS, EEE, NTU Singapore-TUM Germany |PMI-PMP|CSM| Lean Six Sigma Black Belt, USA

River Publishers

Routledge
Taylor & Francis Group
NEW YORK AND LONDON

## RIVER RAPIDS SERIES ON INTELLIGENCE IN CHIPS

The primary goal of the **River Rapids Series on Intelligence in Chips** is to provide a comprehensive collection of instructive books that dive into the varied world of memristors, emerging devices, circuits, systems and investigate their applications in numerous cutting-edge sectors. Memristors and similar memories have gotten a lot of interest because of their unique qualities, which include non-volatility, low power consumption, and the capacity to simulate synapses in neural networks, among other things. The lessons are intended to provide in-depth insights on memristors and their applications in a variety of cutting-edge sectors of computing and engineering.

Topics covered in the series include but are not limited to:

- Memristor Fundamentals
- Neural Networks
- Cellular Neural Networks
- Memristor Array Computing
- Quantum AI Computing
- Efficient Reconfigurable Computing
- Bioinspired Circuits and Systems
- Interdisciplinary Insights
- Practical Implementations
- Emerging Trends

For a list of other books in this series, visit www.riverpublishers.com

# Contents

# Preface

In the dynamic landscape of semiconductor manufacturing, the focus on semiconductors, manufacturing processes, and defect prevention are paramount. Traditional approaches have yielded valuable insights, yet the emergence of artificial intelligence (AI) and Internet of Things (IoT) technologies heralds a new era in defect prevention strategies.

If you are a semiconductor professional seeking to expand your knowledge on silicon processes, delve into the significance of defect prevention, and explore methods for optimizing processes by reducing defects using AI and IoT technologies, this book is for you. Similarly, if you are an engineer specializing in AI and machine learning aiming to transition into the semiconductor industry to contribute to process optimization, this book offers valuable insights. It is also ideal for interdisciplinary researchers engaged in defect analysis, materials science, semiconductor process optimization, and AI-ML applications at the cutting edge of technology. Additionally, early graduates or Master's students aspiring to enter the semiconductor industry or corporate sector will find this book beneficial for gaining foundational knowledge and understanding industry practices.

This meticulously crafted book aims to provide concise yet insightful content tailored to today's fast-paced readers. By emphasizing semiconductors, manufacturing processes, and defect prevention, it offers a comprehensive understanding of these critical areas. Notably, the integration of AI and IoT in chip manufacturing defect prevention represents a groundbreaking advancement.

In my experience, finding integrated content covering these diverse topics has been challenging. This book's format ensures that readers can efficiently grasp key concepts without being overwhelmed by unnecessary length. Targeting semiconductor engineers, researchers, technology professionals, and students, it serves as a valuable resource in understanding the intricate interplay between semiconductors, manufacturing processes, defects, and the transformative potential of AI and IoT integration. The subsequent sections delve into mathematical equations essential for failure analysis and parameter control, providing practical tools for cause analysis, research, and industry professionals. By envisioning hypothetical use cases and exploring theoretical applications, the book inspires innovation and fosters discussions on the transformative impact of AI and IoT integration.

Through interdisciplinary insights from semiconductor engineering, AI, and IoT, this book charts a course toward a future where semiconductor manufacturing defects are minimized, productivity is maximized, and innovation thrives at the intersection of technology and industry. The production process of this book has involved several critical stages, including writing, peer reviews, proofreading, and incorporating feedback, all within a comprehensive quality management system. I have diligently endeavored to ensure the book is free from errors to the best of my ability. However, should any mistakes have inadvertently occurred, I sincerely regret them and would be immensely grateful to anyone who brings them to my attention. Please note that I do not assume legal responsibility, and the author shall not be liable for any direct, indirect, incidental, consequential, or punitive damages arising from the use of or inability to use the content of this book.

May you all be blessed with success. Happy reading!!

# About the book

Welcome to the journey through the intricate world of semiconductor manufacturing. This book aims to provide a comprehensive overview of various aspects of semiconductor fabrication processes, defects, emerging trends, and the integration of cutting-edge technologies like artificial intelligence (AI) and the Internet of Things (IoT) into semiconductor manufacturing.

Chapter 1 is an introduction to semiconductor industry breakdown, including technologies in chip design and development (frontend/pre-silicon) and post-silicon testing. It is drafted to understand the synergy between pre-silicon and post-silicon processes and the differences between fabs, fabless, foundry, IDM, and OSAT companies.

Chapter 2 explores emerging trends in semiconductor manufacturing through case studies, illustrating how these advancements shape the industry's future and an overview of the "shift-left" approach in early defect prevention and process optimization.

Chapter 3 provides detailed insights into wafer fabrication, thin film deposition, photolithography, etching, doping/diffusion of impurities, and oxidation processes. It also introduces the critical role of defect prevention in semiconductor manufacturing and the challenges associated with traditional defect prevention methods.

Chapter 4 focuses on understanding critical dimensions (CD) and factors affecting lithography and etch profiles. It explores mathematical equations governing thin film deposition, photolithography, etching, and doping processes.

Chapter 5 covers different probing techniques, mathematical equations in probing, and an introduction to testing and packaging.

Chapter 6 emphasizes the importance of defect prevention, shift-left in semiconductor manufacturing, and process optimization strategies with case studies.

Chapter 7 incorporates AI/IoT into process optimization, data collection, preprocessing, model development, real-time monitoring, and edge computing.

Chapter 8 integrates IoT technologies in semiconductor manufacturing for smart manufacturing, predictive maintenance, quality control, supply chain management, and real-time process control.

Chapter 9 delves into the fundamentals of semiconductor defects, examining types, sources, effects, detection, and characterization of defects in semiconductor manufacturing, along with hypothetical case studies addressing specific defect scenarios in various processes. It also analyzes the challenges of traditional defect prevention methods and highlights the need for proactive, data-driven defect prevention approaches. This chapter explores emerging trends in semiconductor manufacturing through case studies, illustrating how these advancements shape the industry's future. A summary and key findings conclude the book, aiming to equip engineers, researchers, and professionals involved in semiconductor design, fabrication, and quality control with a comprehensive understanding of defect prevention strategies.

# About the Author

**Rupal Jain** brings a wealth of experience in engineering, program management, and strategic alignment, and stands as an eminent personality in the realm of semiconductor chip manufacturing. She has orchestrated projects spanning the entire spectrum from design conception and quality management to final delivery on a global scale, covering regions such as the USA, Taiwan, Singapore, Italy, Malaysia, China, and India. Recognized for her profound expertise, Rupal is celebrated by prestigious certifications such as PMP, CSM, and Lean Six Sigma Black Belt. She holds a Master's in Electrical and Electronics Engineering from NTU Singapore in a joint degree program with TUM Germany. Her contributions have garnered international acclaim, earning her invitations to esteemed publications, jury, nominated memberships, and coveted awards. Her other authored pieces, "Semiconductor Essentials: A Leader's Express Reference to Electronics Concepts" and "Mastering Project Management: PMP and Agile for Leaders", promise to share her valuable insights with the next generation of leaders and engineers.

CHAPTER

# 1

# Introduction

Semiconductors [1] form the foundation of modern electronics, serving as the building blocks for a wide range of electronic devices that have revolutionized our lives. These materials, with unique electrical properties lying between conductors and insulators, play a crucial role in enabling the functionality of electronic circuits, from microchips in smartphones to complex integrated circuits in computers.

In this chapter we will delve into an introduction to semiconductors, exploring their evolution, providing an overview, and introducing various topics such as process steps, equations, and pre-silicon, and post-silicon strategies. These discussions aim to offer a concise yet thorough understanding of equations and process steps. Additionally, later sections in this chapter will introduce the definitions of AI [20] and IoT, along with their integration with semiconductor process technologies.

At the heart of semiconductor technology lies the manipulation of electron behavior within semiconductor materials, typically silicon. By carefully controlling the movement and behavior of electrons through processes like doping, etching, and deposition, semiconductor manufacturers can create intricate circuits capable of performing a myriad of functions.

Semiconductors are indispensable in today's technological landscape, powering devices and systems across various industries, including telecommunications, automotive, healthcare, and consumer electronics. The relentless pursuit of miniaturization, efficiency, and performance improvement continues to drive innovation in semiconductor manufacturing, paving the way for the development of increasingly powerful and sophisticated electronic devices. But did you know that its journey starts from something as simple as sand.

**From Humble Sand to Mighty Chips: The Semiconductor Journey**

Semiconductors, the brains behind modern electronics, might surprise you with their origin story. Their journey begins not in a high-tech lab, but on a seemingly ordinary beach – with sand!

Sand is a loose granular material composed of tiny rock and mineral fragments. These fragments come from the erosion of larger rocks by wind, water, or glaciers. The most common component of sand is silicon dioxide ($SiO_2$), also known as silica [2].

**Figure 1.1:** Sand to Silicon: Introduction.

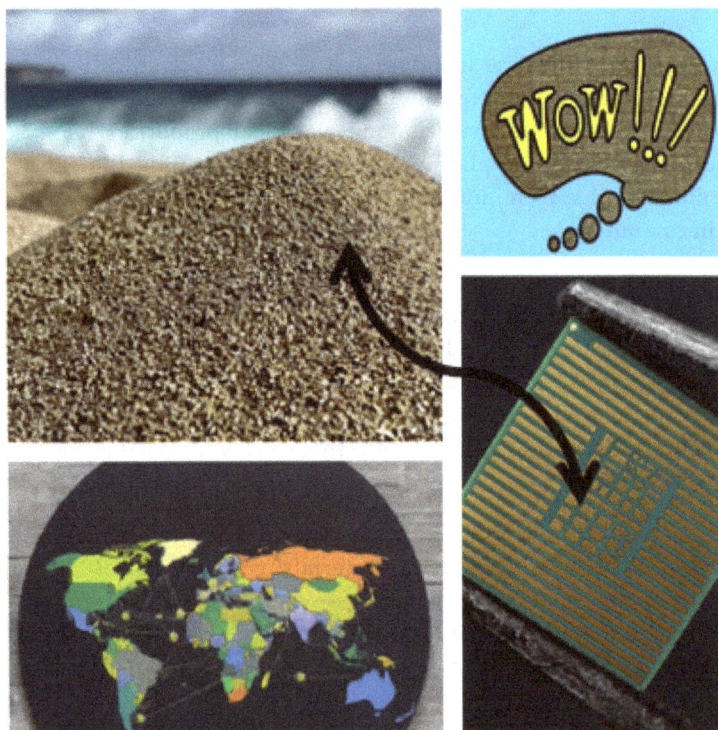

**Figure 1.2:** Semiconductor industry applications.

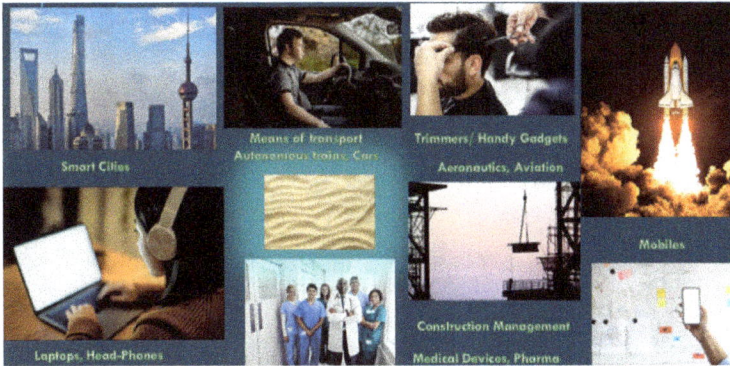

**Figure 1.3:** Timeline and evolution of components that shaped industry.

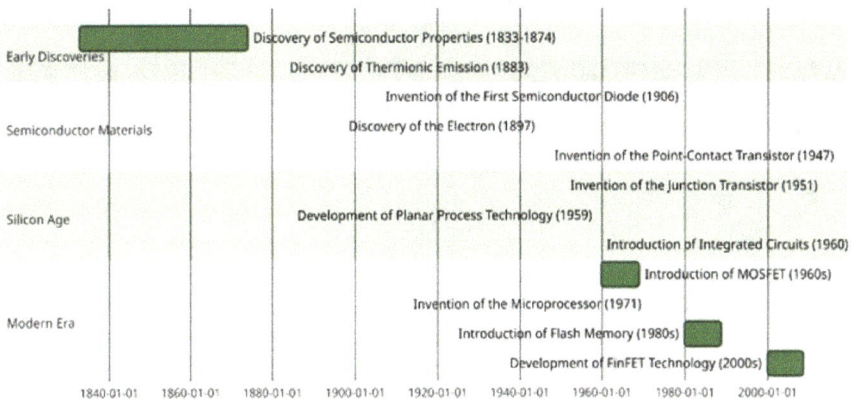

## Discovery of electrons [1, 2] (late 19th century)

- J.J. Thomson's discovery of the electron in 1897 laid the groundwork for understanding the behavior of electrons in materials.

## Invention of the vacuum tube [1] (1904–1906)

- John Ambrose Fleming's development of the vacuum tube diode (1904) and Lee De Forest's invention of the triode (1906) marked significant milestones in electronic device technology [11].

## Semiconductor discoveries [1] (1920s–1940s)

- Early research on semiconductors began with the discovery of the semiconductor properties of materials like germanium and silicon in the 1920s and 1930s.
- In 1947, Bell Labs scientists John Bardeen, Walter Brattain, and William Shockley invented the point-contact transistor, leading to the birth of solid-state electronics.

## Transistor era [1] (1950s)

- The invention of the bipolar junction transistor (BJT) in 1947 and the development of the first silicon transistor in 1954 by Morris Tanenbaum and John Pearson marked the beginning of the transistor era.
- The invention of the integrated circuit (IC) by Jack Kilby at Texas Instruments in 1958 revolutionized electronics by enabling multiple transistors to be fabricated on a single semiconductor substrate.

## Advancements in Semiconductor manufacturing [2] (1960s–1970s)

- The 1960s saw significant advancements in semiconductor manufacturing processes, including the development of planar technology by Jean Hoerni at Fairchild Semiconductor.
- The introduction of MOS (metal-Oxide-Semiconductor) technology in the late 1960s paved the way for the fabrication of smaller and more efficient transistors.
- The microprocessor, a complete central processing unit (CPU) on a single chip, was invented by Intel in 1971, marking a major milestone in computing history.

## Rapid miniaturization [1] (1980s–1990s)

- The 1980s and 1990s witnessed rapid miniaturization of semiconductor devices, driven by advancements in photolithography, materials science, and process technology.
- Introduction of complementary metal-oxide-semiconductor (CMOS) technology enabled lower power consumption and greater integration density.
- The development of flash memory in the 1980s and 1990s revolutionized data storage, leading to the widespread adoption of solid-state storage devices.
- The emergence of fabless semiconductor companies and the globalization of semiconductor manufacturing further accelerated innovation and competition in the industry.

## Internet age and digital revolution [1] (late 1990s)

- The late 1990s saw the rise of the internet age and the digital revolution, fueled by advancements in semiconductor technology.
- The dot-com boom led to increased demand for semiconductor devices used in telecommunications, networking, and consumer electronics.

A brief description of the process is shown in the flow chart of Figure 1.4.

**Figure 1.4:** Process flowchart.

**The early steps: Extracting silicon [12]:** Our journey starts with extracting silicon, the key ingredient for semiconductors, from sand. However, not just any sand will do. The industry uses a specific type called silica sand, which has a high concentration of silicon dioxide (around 95%). This sand is extracted from quarries using various methods like open-pit mining or dredging.

**Purification: From sand to silicon [1]:** The extracted sand isn't pure silicon yet. It contains impurities like iron and aluminum that can disrupt the electrical properties needed for semiconductors. Here's where the magic happens (Figure 1.5):

**Figure 1.5:** Purification process.

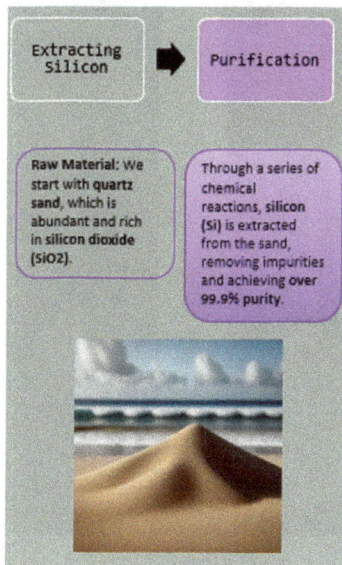

- **Heating and reacting:** The sand is heated to extremely high temperatures and mixed with a reducing agent like carbon. This process removes oxygen from the silicon dioxide, leaving behind a purer form of silicon called metallurgical grade silicon (MG-Si).
- **Further refining:** This MG-Si still contains some impurities. Additional chemical processes are used to achieve the ultra-pure electronic grade silicon (EG-Si) needed for semiconductors. This EG-Si is at least 99.9999% pure!

**Crystallography [2]: Creating the perfect silicon structure:** Semiconductors require silicon with a specific crystal structure for optimal performance. Here's how they achieve it:

- **The seed crystal:** A small, perfectly formed silicon crystal (seed) is used as the starting point.
- **The melt:** Molten silicon [12] (liquid silicon) is prepared at high temperatures.
- **The pulling process:** The seed crystal is slowly pulled upwards from the molten silicon. As it rises, silicon atoms from the melt arrange themselves in the same crystal structure as the seed, forming a large, single-crystal silicon ingot (Figure 1.6).

**Figure 1.6:** Crystallization, ingot.

**Slicing [2] and dicing: Wafers for chip creation:** The large silicon ingot is then sliced into incredibly thin wafers (Figure 1.7) using diamond saws. These wafers, typically around 0.02 millimeters thick, become the foundation for creating integrated circuits (ICs) or microchips.

**Figure 1.7:** Bare silicon wafer.

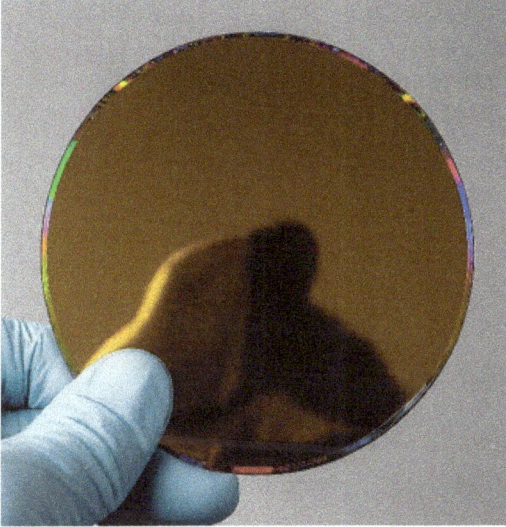

*Fun Fact: Do you know why wafers are always circular in shape?*

*Answer: When the CZ process is performed, the ingot is cylindrical and cutting in the shape of circle when ingot is sliced horizontally is easier. This shows effective material utilization and process.*

**The final frontier: From wafers [12] to chips:** The journey continues as these wafers undergo a complex series of photolithography, etching, and doping processes to create the intricate circuits that power our devices. This is a whole new adventure in itself, but for now, we can appreciate the remarkable transformation of humble sand into the foundation of modern technology.

We'll delve into the process in more depth in upcoming sections. We'll explain how bare silicon wafers receive prints or designs in simple terms, covering topics like reticles, steps involved, and more. Additionally, we'll explore the equations that underpin these processes. These equations are essential for professionals in industry and are also crucial for higher education and research purposes [12].

**Figure 1.8:** Process Flow representing the journey from Bare Silicon to Integrated chips.

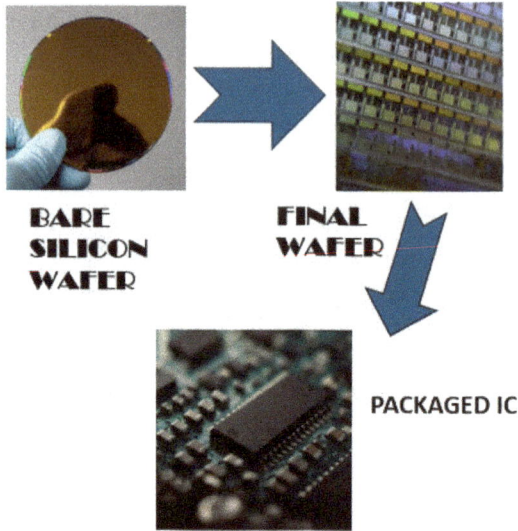

BARE
SILICON
WAFER

FINAL
WAFER

PACKAGED IC

## 1.1 Breakdown of the Semiconductor Industry

The Semiconductor industry is very dynamic. The innovation and operations span from design to development, followed by operations.

There is a front-end part to the breakdown and a back-end part (Figure 1.9). In the world of semiconductor manufacturing, the journey from concept to finished chip involves two key stages: *pre-silicon* and *post-silicon*. These stages

**Figure 1.9:** Breakdown of the semiconductor industry.

Front-end

•Pre-silicon/ Fabless

Backend

•Post silicon (Fabs, Foundry, IDM, OSAT)

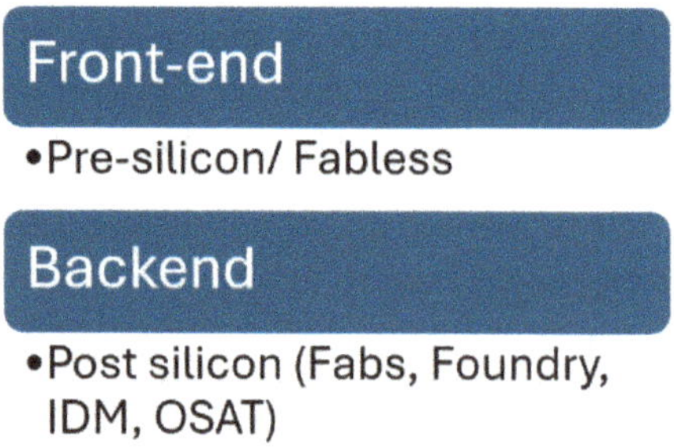

represent distinct phases in testing and verification, each crucial for ensuring a functional and reliable final product.

## 1.1.1 Technologies in chip design and development (frontend/pre-silicon)

Imagine building a complex machine entirely in a computer simulation. That's essentially what pre-silicon validation is all about. Here, engineers leverage sophisticated tools to meticulously examine the chip's design before any physical silicon is manufactured (Figure 1.10) [67].

- **Techniques:**

  - o **Simulation:** Virtual models of the chip are created and subjected to various test cases, mimicking real-world scenarios. This helps identify potential design flaws like timing issues or logical errors.
  - o **Emulation:** Software emulators act as a stand-in for the actual chip, allowing engineers to run real code and observe behavior in a controlled environment. This provides a more realistic test compared to simulation.
  - o **Formal verification [14]:** This technique uses mathematical proofs to formally analyze the chip's design for adherence to specifications. It helps catch errors in logic and ensures the design fulfills its intended purpose.

**Figure 1.10:** Hypothetical view of frontend of semiconductor design and development.

- **Example:** Let's say you're designing a new microprocessor. Pre-silicon simulation can reveal issues like data corruption due to incorrect timing margins or unexpected interactions between different components. Addressing these issues virtually is far more efficient than discovering them after physical chips are manufactured.

The front end of chip design (Figure 1.11) encompasses several key technologies that define the functionality and structure of integrated circuits (ICs). Here are some of the most prominent:

**Figure 1.11:** Frontend tools in design.

1. **Hardware description languages [14] (HDLs)**

   - **Verilog and VHDL:** These are the primary languages used to describe the digital logic of an IC using a text-based format. They allow designers to define the behavior and functionality of the circuit at a high level.

2. **Computer-aided design (CAD) tools**

   - **EDA [14] (electronic design automation) tools:** These software applications assist designers in various stages of the design process, including:

     - **Schematic capture:** Visually representing the circuit using graphical symbols and connections.
     - **Simulation:** Verifying the functionality of the designed circuit under different conditions using simulation tools.
     - **Synthesis:** Converting the HDL code into an optimized netlist, a low-level representation of the circuit's building blocks.
     - **Place and route:** Physically arranging the circuit components (cells) on the chip and defining the connections between them.

3. **Design for testability (DFT)**

- This methodology incorporates features into the design phase that enable efficient testing of the manufactured chip to identify and troubleshoot any potential defects.
- **Common DFT [66] techniques include:**

  o **Scan insertion:** Adding temporary circuitry to allow for easier control and observation of internal nodes within the chip during testing.
  o **Boundary scan (BST):** Enabling testing of connections between the chip and external pins.
  o **Built-in self-test (BIST):** Incorporating dedicated circuitry within the chip to perform self-testing upon power-up.

4. **Physical design and verification**

- **Layout:** Arranging and connecting the various components like transistors, wires, and routing paths on the chip layout using specialized tools.
- **Static timing analysis [66] (STA):** Analyzing the timing characteristics of the chip to ensure signals propagate within the required time constraints.
- **Layout versus schematic (LVS) verification:** Comparing the physical layout of the chip with the original HDL [67] code to ensure they represent the same functionality.

5. **Foundry-specific design rules**

- Each chip manufacturer (foundry) has its own set of design rules that define the limitations and specifications for designing circuits with their specific fabrication processes. Adhering to these rules is crucial for ensuring the manufacturability and functionality of the chip.

**Industries utilizing chip design and development:**

- **Semiconductor industry:** This sector designs and manufactures various types of ICs, including microprocessors, memory chips, and application-specific integrated circuits (ASICs), for various applications.
- **Consumer electronics:** The electronics industry uses chips in a wide range of products, from smartphones and laptops to televisions and gaming consoles.
- **Automotive industry:** Modern vehicles rely heavily on advanced electronics and require specialized chips for engine control, infotainment systems, and driver-assistance features.

- **Communication and networking:** The infrastructure for communication networks, including routers, switches, and base stations, utilizes various types of chips for data processing and transmission.
- **Aerospace and defense:** This industry utilizes highly specialized and reliable chips for applications like satellite communication, navigation systems, and weapons systems.
- **Medical devices:** Medical devices like pacemakers, hearing aids, and diagnostic equipment often incorporate specialized chips to perform specific functions.

**Figure 1.12:** Substrate mapping after floor planning view on PC.

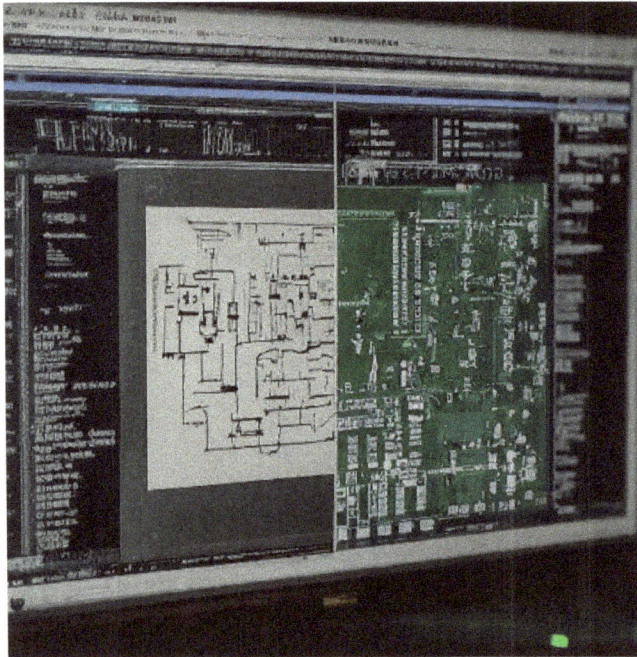

### 1.1.1.1 Tools of the trade: CAD, EDA, simulation, and synthesis

The world of chip design is a complex dance between creativity and precision. To navigate this intricate process, engineers rely on a powerful arsenal of tools [60]:

- **Computer-aided design (CAD):** This broad term encompasses any software that assists in creating designs. In the realm of chip design, specialized electronic design automation (EDA) tools come into play [60, 68].

- **Electronic design automation (EDA):** A suite of software applications dedicated to the lifecycle of an integrated circuit (IC) design. EDA tools encompass various functions, including [60, 68]:

  - **Schematic capture:** Tools like Cadence Virtuoso or Synopsys Custom Compiler allow engineers to create a visual representation of the circuit using symbols and connections.
  - **Hardware description languages (HDLs):** Languages like VerilogVerilog or VHDL offer a text-based approach for describing the chip's functionality. Tools like Cadence Incisive or Synopsys Verdi help with HDL coding and verification.
  - **Simulation:** Software emulators like Cadence Virtuoso Simulator or Synopsys VCS mimic the chip's behavior, allowing engineers to test functionality under various scenarios.
  - **Place and route (PnR):** Tools like Cadence Innovus or Synopsys IC Compiler determine the physical placement of components on the chip and route connections between them.
  - **Layout verification:** Ensures the physical design adheres to manufacturing rules. Tools like Cadence Dracula or Synopsys Calibre perform these checks.

- **Synthesis:** This process transforms a high-level HDL description into a netlist, a low-level representation of the circuit suitable for place and route tools. Synopsys Design Compiler and Cadence Genus are popular synthesis tools.

#### 1.1.1.2 Case Study: FPGA design verification using Xilinx Vivado.

Field-programmable gate arrays (FPGAs) offer a unique design paradigm. Unlike custom chips, FPGAs consist of pre-defined logic blocks that can be configured to implement various functionalities. Xilinx Vivado [61] Design Suite is a powerful EDA toolchain specifically tailored for FPGA design using Xilinx devices [61, 68, 69, 70].

**Scenario:** An engineer is designing a digital signal processing (DSP) filter using a Xilinx FPGA. Here's how Vivado aids in verification:

1. **HDL coding and simulation:** The engineer uses Vivado's built-in HDL editor to create a Verilog description of the DSP filter. Vivado Simulator allows comprehensive functional simulation of the design, stimulating it with various test vectors to ensure correct filtering behavior.
2. **Synthesis and implementation:** Vivado's synthesis engine translates the Verilog code into a netlist optimized for the target Xilinx FPGA [61]. The implementation phase performs place and route, mapping the netlist onto the FPGA fabric and creating the bitstream configuration file.

3. **Static timing analysis (STA):** Vivado's timing analysis tools ensure that signals propagate through the FPGA fabric within specified timing constraints. This helps prevent glitches or errors caused by slow signal propagation [61].

4. **Post-place and route simulation (place and route (PnR) simulation):** Vivado allows simulating the design after place and route, incorporating the actual delays introduced by the FPGA's routing fabric. This provides a more realistic picture of the design's behavior on the actual hardware.

5. **Hardware co-simulation:** For complex designs, Vivado [61] enables co-simulation, where the FPGA design interacts with a hardware model (e.g., a microcontroller) simulated in software. This allows for comprehensive system-level verification.

**Benefits of using Xilinx Vivado for FPGA verification:**

- **Integrated toolchain:** Vivado [61] offers a cohesive environment for all stages of FPGA design verification, streamlining the workflow.
- **Device-specific optimization:** The tools are specifically designed for Xilinx FPGAs, ensuring optimal performance and accurate results.
- **Comprehensive verification features:** Vivado [61] provides a rich set of features for simulation, timing analysis, and co-simulation, enabling thorough design validation.

### 1.1.2 Post-silicon: Testing in the real world

Once the pre-silicon phase is complete, it's time to validate the design on actual silicon. This is where the chips, manufactured in a fab (fabrication facility), come into play.

### 1.1.2.1 Techniques

**Functional testing:** Actual chips are mounted on test boards and bombarded with pre-defined test cases to verify their functionality against design specifications. This ensures the chip performs its intended operations correctly.

**Performance testing:** Engineers measure the chip's performance metrics like speed, power consumption, and heat generation. This helps ensure the chip meets the performance targets set during the design phase.

**Debug and analysis:** If any issues arise during testing, engineers use sophisticated tools to analyze the chip's behavior and identify the root cause of the problem. This may lead to design modifications or bug fixes.

**Case study:** Remember the infamous Pentium FDIV [69] bug? This mathematical error in the chip's floating-point unit went undetected during pre-silicon verification. It was only discovered during post-silicon testing, leading to a major recall and reputational damage for Intel. This case highlights the importance of thorough testing at both stages.

## 1.1.2.2 Types of testing

### 1. Automated test equipment (ATE):

- This refers to a sophisticated system that automatically performs a series of electrical tests on integrated circuits (ICs) after they are manufactured and singulated from the wafer.
- ATE uses computer-controlled hardware and software to stimulate the IC with various test patterns and analyze its responses, comparing them to pre-defined specifications.
- This testing helps identify defects like shorts, opens, and functional errors in the IC, enabling manufacturers to quickly identify and discard non-functional devices [70].

### 2. Scan-based testing (SLT):

- This technique utilizes built-in test structures (scan chains) embedded within the IC design for efficient testing.
- Scan chains are essentially "bypass routes" that allow testers to control and observe the internal state of the IC by shifting data through them.
- SLT [70] enables testing of logic blocks, memory elements, and interconnects within the IC, providing valuable information beyond the basic functionality verified by ATE.
- Compared to ATE, SLT can offer higher test coverage and detect smaller defects, especially in highly complex ICs.

### 3. Additional Testing Techniques

- **Boundary scan test (BST):** This method uses dedicated circuitry embedded within the IC to test the connections (boundaries) between the IC and external pins [70].
- **In-circuit test (ICT):** This testing occurs after the IC is already assembled onto a printed circuit board (PCB) and checks the functionality of the entire circuit board, including the interaction between the IC and other components [70].
- **Burn-in test:** This involves stressing the IC at elevated temperatures and voltages for an extended period to identify potential weaknesses or latent defects that might not be apparent under normal operating conditions [69, 70].

### 1.1.3 The synergy of pre-silicon and post-silicon

While distinct, pre-silicon and post-silicon validation work hand-in-hand, pre-silicon helps catch most design errors early, saving time and resources. However, post-silicon validation is crucial for uncovering issues that might not be apparent in a virtual environment. The ideal scenario is to minimize post-silicon surprises by achieving a high degree of confidence in the design through comprehensive pre-silicon verification.

The semiconductor industry is constantly pushing boundaries. With ever-increasing chip complexity, the focus is on improving pre-silicon verification methodologies to catch more issues upfront. Techniques like "shift-left" verification aim to move some aspects of post-silicon validation earlier in the design cycle, further optimizing the overall process.

By understanding the roles of pre-silicon and post-silicon validation, we gain a deeper appreciation for the intricate process of bringing complex chips to life. It's a testament to the ingenuity of engineers who strive for flawless silicon, laying the foundation for the countless electronic devices that power our modern world [70].

#### 1.1.3.1 Rewards

While the challenges in chip manufacturing are significant, successfully overcoming them brings a multitude of rewards, not just for the industry itself, but for society as a whole. Here are some of the key rewards:

**1. Technological advancement:**

- **Moore's law:** Continued advancements in chip manufacturing allow the industry to uphold Moore's [1] law, which predicts the doubling of transistors on a chip roughly every two years. This exponential growth in computing power fuels innovation across diverse fields like artificial intelligence, healthcare, robotics, and more.
- **Smaller, faster, and more efficient devices:** Smaller and more efficient chips enable the development of smaller, lighter, and more energy-efficient electronic devices. This benefits consumers with wider device options, longer battery life, and reduced environmental impact.

**2. Economic growth:**

- **Job creation:** The chip manufacturing industry is a major job creator, employing millions of people worldwide in various roles like engineering, manufacturing, and research.

- **Economic competitiveness:** Leading the way in chip development and production strengthens a nation's economic competitiveness by fostering innovation, attracting investments, and supporting various technology-driven industries.

## 3. Improved quality of life:

- **Technological innovations:** Advancements in chip technology have a profound impact on almost every aspect of our lives. From enabling powerful mobile devices and communication networks to supporting medical advancements and life-saving technologies, these innovations improve the quality of life for individuals and communities.
- **Increased access to information and communication:** Efficient and affordable chips power the devices that connect us to information, education, and global communication networks. This access fosters knowledge sharing, collaboration, and promotes positive social change.

### 1.1.3.2 Pushing the boundaries of science and engineering and effective technology management

Overcoming the challenges in chip manufacturing requires constant innovation in materials science and physics. This fosters research and development that pushes the boundaries of knowledge and understanding, leading to potential breakthroughs in diverse scientific fields. The quest for smaller and more powerful chips leads to the development of entirely new technologies and applications. This continuous evolution keeps pushing the frontiers of what's possible, paving the way for groundbreaking advancements in the future.

In addition to the rewards directly tied to overcoming chip manufacturing challenges, effective technology management offers a wider range of benefits for businesses and organizations:

1. **Enhanced efficiency and productivity:** Implementing efficient processes, automation, and data-driven decision-making can streamline operations, improve resource utilization, and ultimately lead to increased productivity and output.
2. **Improved risk management:** Proactive technology management involves anticipating potential risks, such as security vulnerabilities or outdated technology, and implementing preventative measures to mitigate these risks before they disrupt operations.
3. **Competitive advantage:** By strategically utilizing technology to optimize processes, improve product offerings, and enhance customer experience, businesses can gain a competitive edge in their respective markets.
4. **Increased innovation:** Effective technology management fosters a culture of innovation by encouraging experimentation with new technologies, exploring creative solutions, and nurturing collaboration between teams.

5. **Improved decision-making:** By leveraging data analytics and insights gained from technology usage, organizations can make informed decisions based on evidence rather than intuition, leading to better long-term results.
6. **Increased customer satisfaction:** By deploying technology to enhance customer service capabilities, provide personalized experiences, and address customer needs efficiently, businesses can improve customer satisfaction and loyalty.
7. **Environmental sustainability:** Implementing sustainable practices within technology management, such as energy-efficient equipment and responsible waste disposal, can reduce an organization's environmental footprint.
8. **Improved employee engagement:** Utilizing technology to provide employees with user-friendly tools, efficient workflows, and access to knowledge bases can enhance their work experience and foster greater engagement.

### 1.1.4 Difference between fabs, fabless companies, foundries, IDMs

1. Fabs (fabrication facilities):

   - Fabs are manufacturing facilities where semiconductor devices are fabricated on silicon wafers through a series of complex processes, including lithography, etching, doping, and metallization [62].
   - These facilities are typically owned and operated by semiconductor companies or foundries and are equipped with specialized equipment for the production of integrated circuits (ICs) and other semiconductor products [67].
   - Fabs can be categorized into different types based on their technology nodes, production capacity, and specialization, such as logic fabs, memory fabs, and analog fabs.
   - Examples of companies with fabs include Intel, TSMC, Samsung, and Global-Foundries.

2. Foundries:

   - Foundries are specialized semiconductor manufacturing companies that offer fabrication services to fabless companies and IDM [62].
   - Unlike IDM, foundries do not design or own the intellectual property (IP) for semiconductor products but provide manufacturing capabilities and expertise to produce ICs based on designs provided by their customers.
   - Foundries typically offer a wide range of process technologies and production capacities to accommodate diverse customer requirements.
   - Examples of leading foundries include TSMC, GlobalFoundries, UMC (United Microelectronics Corporation), and SMIC (Semiconductor Manufacturing International Corporation) [62, 63, 64, 67].

3.  Fabless companies:

    - Fabless companies are semiconductor firms that focus solely on the design and development of semiconductor products without owning semiconductor fabrication facilities (fabs) [62, 63, 64, 67].
    - These companies specialize in chip design, including integrated circuits (ICs), system-on-chip (SoC) solutions, and other semiconductor devices, leveraging third-party foundries for manufacturing.
    - Fabless companies benefit from lower capital investment and greater flexibility by outsourcing fabrication to foundries, allowing them to focus on innovation and product differentiation.
    - Examples of fabless companies include Qualcomm [67], NVIDIA, AMD (Advanced Micro Devices), and Broadcom [62].

4.  IDM (integrated device manufacturer):

    - IDM refers to semiconductor companies that both design and manufacture their semiconductor products in-house, owning both the intellectual property (IP) and fabrication facilities (fabs).
    - These companies have complete control over the entire semiconductor manufacturing process, from design to fabrication, enabling tighter integration and optimization of products [62].
    - IDM typically invests heavily in research and development (R&D) to develop advanced semiconductor technologies and maintain competitiveness in the market.
    - Examples of IDM include Intel, Samsung Electronics, Texas Instruments, and Micron Technology.

5.  OSAT (outsourced semiconductor assembly and test) companies:

    - OSAT companies provide semiconductor packaging and testing services to semiconductor manufacturers, fabless companies, and IDM [62].
    - These companies specialize in assembling individual semiconductor components into packaged ICs, as well as testing the functionality and reliability of the finished products.
    - OSAT companies offer a wide range of packaging technologies, including flip-chip, wire bonding, and through-silicon vias [13] (TSVs), to meet diverse customer requirements.
    - Examples of OSAT companies include ASE Technology [62] Holding, Amkor Technology, JCET (Jiangsu Changjiang Electronics Technology), and Powertech Technology Inc. [67].

## 1.1.4.1 Some examples of companies involved in silicon manufacturing

### Fabs:

**Intel Corporation:** Known for its advanced semiconductor manufacturing facilities, Intel operates multiple fabs worldwide, producing a wide range of microprocessors and other semiconductor products [47].

**Taiwan Semiconductor Manufacturing Company (TSMC):** TSMC is the world's largest dedicated independent semiconductor foundry, operating several advanced fabs in Taiwan and other locations globally [52].

### Fabless companies:

**Qualcomm [63] Incorporated:** As a leading fabless semiconductor company, Qualcomm designs and develops cutting-edge system-on-chip (SoC) solutions for mobile devices, automotive, IoT, and other applications.

**NVIDIA [52] Corporation:** NVIDIA is renowned for its innovative graphics processing units (GPUs) and AIAI accelerators, designed by its team of engineers without owning semiconductor fabrication facilities.

### Foundries:

**GlobalFoundries:** GlobalFoundries is a leading [54] semiconductor foundry [55] offering advanced manufacturing solutions for a wide range of customers, including fabless companies and IDM [53].

**Samsung Foundry:** Samsung Foundry provides advanced semiconductor manufacturing services utilizing Samsung's state-of-the-art fabs [56].

### IDM (integrated device Integrated Device Manufacturer):

**Samsung Electronics:** Samsung is a prominent IDM that designs, manufactures, and sells a diverse range of semiconductor products, including memory chips, logic chips, and application processors [56].

**Texas Instruments [65] (TI):** TI is an example of an IDM that designs and manufactures analog and digital semiconductor products for various industries, including automotive, industrial, and consumer electronics.

## OSAT (outsourced semiconductor assembly and test) companies:

**ASE [62, 67] Technology Holding Co., Ltd.:** ASE is one of the world's largest OSAT companies, offering semiconductor packaging and testing services for a broad range of applications.

**Amkor Technology, Inc.:** Amkor [64] is a leading provider of semiconductor packaging and test services, offering advanced solutions to fabless companies, IDM, and foundry customers.

# 2

# Exploring Emerging Trends in Semiconductor Manufacturing

When discussing a 3000 mm wafer in semiconductor fabrication, we are addressing the dimensions of the silicon wafer utilized as the foundational substrate for manufacturing integrated circuits (ICs) and other semiconductor devices. The designation "3000 mm" typically denotes the diameter of the wafer.

In the semiconductor realm, the prevalent wafer sizes include 200 mm (8 inches), 300 mm (12 inches), and more recently, 450 mm (18 inches). Larger wafer dimensions, such as 300 mm, are employed to enhance the quantity of chips that can be produced on a single wafer, thereby enhancing manufacturing efficiency and diminishing production costs per chip.

The transition from smaller wafer sizes (e.g., 200 mm) to larger ones (e.g., 300 mm) has been a notable trend in semiconductor manufacturing. Larger wafers enable the simultaneous fabrication of more chips, leading to economies of scale and heightened yields [71]. However, the shift to even larger wafer sizes, such as 450 mm, has encountered greater challenges due to technological and economic factors. Consequently, as chip quantities increase, there is a requisite for effective quality assessments to ensure defect-free chip manufacturing.

It is important to acknowledge that the wafer size can significantly influence the overall production process, equipment, and cost structure within a semiconductor fabrication facility. The selection of wafer size is influenced by factors such as cost-effectiveness, manufacturing efficiency, and the technological capabilities of the semiconductor industry at a given juncture.

The pitch size and resolution in semiconductor manufacturing are closely related parameters that impact the quality, precision, and functionality of integrated circuits (ICs) and semiconductor devices. Here's how they are affected and some methods to achieve the best resolution [3]:

1. **Effect of pitch size on resolution**

   - The pitch size refers to the distance between identical features on a semiconductor device, such as transistors, interconnects, or memory cells.
   - Smaller pitch sizes allow for denser packing of features on the semiconductor device, enabling higher integration levels and improved performance.
   - However, reducing the pitch size poses challenges in terms of lithography, alignment, and manufacturing tolerances, which can affect the achievable resolution [3].

2. **Effect of resolution on device performance**

   - Resolution refers to the smallest feature size that can be accurately reproduced during the lithography process.
   - Higher resolution enables the fabrication of finer details and smaller structures on the semiconductor device, leading to improved device performance, speed, and power efficiency.
   - Achieving higher resolution is crucial for advanced semiconductor technologies, such as nanoscale process nodes, where feature sizes continue to shrink to meet the demands of Moore's law.

3. **Methods to achieve best resolution**

   - **Advanced lithography techniques:** Utilizing advanced lithography [3] techniques, such as immersion lithography, extreme ultraviolet (EUV) lithography, and multi-patterning, enables the fabrication of smaller features and higher resolution.
   - **Optical enhancement techniques:** Employing optical enhancement techniques, such as phase-shift masks, optical proximity correction (OPC), and resolution enhancement techniques (RET [3]), enhances the resolution and pattern fidelity during the lithography process.
   - **High-resolution metrology:** Implementing high-resolution metrology tools and inspection systems enables accurate measurement and characterization of features on semiconductor devices, ensuring compliance with design specifications and quality standards.
   - **Process optimization:** Optimizing semiconductor manufacturing processes, including photoresist formulation, etching, deposition, and annealing, can improve resolution by minimizing process variations, defects, and material interactions that can degrade pattern fidelity.

- **Materials innovation:** Developing novel materials with improved optical properties, photoresist sensitivity, and etch resistance can enhance lithographic resolution and enable the fabrication of smaller features on semiconductor devices.
- **Mask design and mask manufacturing:** Designing optimized masks with precise feature patterns and utilizing advanced mask manufacturing techniques, such as electron beam lithography and nanoimprint lithography, contribute to achieving higher resolution and pattern fidelity during lithography [3].

## 2.1 Case Studies based on some of the Emerging Trends

1. **ASML**

   - **Technology:** ASML [10] is a leading provider of photolithography equipment, including immersion lithography and extreme ultraviolet (EUV) lithography systems [47].
   - **Case study:** ASML's EUV lithography systems are used by semiconductor manufacturers to produce advanced chips with smaller feature sizes and higher resolution. For example, ASML's EUV systems have been adopted by leading semiconductor manufacturers like TSMC, Samsung, and Intel for their most advanced process nodes, enabling the production of cutting-edge chips for applications such as high-performance computing, artificial intelligence, and 5G.

2. **Nikon Corporation**

   - **Technology:** Nikon Corporation offers a range of lithography equipment, including immersion lithography systems and advanced metrology solutions [48].
   - **Case study:** Nikon's lithography systems are used for high-resolution patterning in semiconductor manufacturing. For instance, Nikon's immersion lithography systems have been utilized by semiconductor manufacturers to achieve finer feature sizes and improved pattern fidelity in their semiconductor devices. Nikon also provides metrology tools for accurate measurement and characterization of semiconductor features, contributing to process optimization and quality control.

3. **KLA Corporation**

   - **Technology:** KLA Corporation specializes in advanced metrology, inspection, and process control solutions for semiconductor manufacturing [49].

- **Case study:** KLA's metrology and inspection tools are used by semiconductor manufacturers to ensure the quality, reliability, and performance of their semiconductor devices. For example, KLA's optical and e-beam inspection systems help detect defects and deviations in semiconductor wafers, enabling early identification and resolution of process issues. KLA's metrology solutions provide precise measurement capabilities for critical dimensions, overlay, and film thickness, supporting process optimization and yield [71] enhancement.

4. **Lam Research Corporation**

- **Technology:** Lam Research offers a range of semiconductor equipment and solutions, including plasma etch systems, deposition systems, and process control software [50].
- **Case study:** Lam Research's plasma etch systems play a critical role in semiconductor manufacturing for pattern transfer and device structuring. For example, Lam's advanced etch systems enable precise etching of semiconductor materials with nanoscale resolution, supporting the fabrication of complex device structures and feature sizes. Lam's process control software provides real-time monitoring and optimization capabilities, ensuring uniformity, repeatability, and yield [71] improvement in semiconductor production.

5. **Applied Materials, Inc.**

- **Technology:** Applied Materials develops semiconductor manufacturing equipment, including deposition systems, etch systems, chemical-mechanical planarization (CMP) systems, and process control solutions [51].
- **Case study:** Applied Materials' deposition and etch systems are essential for depositing thin films and patterning semiconductor materials with high precision and resolution. For instance, Applied Materials' atomic layer deposition (ALD) systems enable the deposition of ultra-thin films with atomic-scale control, supporting advanced semiconductor fabrication processes. Applied Materials' CMP systems provide planarization capabilities for smoothing semiconductor surfaces, ensuring uniformity and reliability in semiconductor device manufacturing.

6. **Intel Corporation**

- **Technology:** Intel is a leading semiconductor manufacturer known for its advanced process technologies and manufacturing capabilities [47].

- **Case study:** Intel's development of 7 nm process technology represents a significant milestone in semiconductor manufacturing. By leveraging advanced lithography techniques, materials innovation, and process optimization, Intel aims to deliver higher performance, energy efficiency, and transistor density in its semiconductor products. The adoption of EUV lithography and other advanced manufacturing techniques is crucial for achieving the desired feature sizes and resolution in Intel's next-generation processors and semiconductor devices.

7.  **TSMC (Taiwan Semiconductor Manufacturing Company)**

- **Technology:** TSMC is the world's largest dedicated semiconductor foundry, specializing in advanced process nodes and manufacturing technologies [52].
- **Case study:** TSMC's development and deployment of 5nm process technology demonstrates its commitment to advancing semiconductor manufacturing capabilities. By investing in EUV lithography, multi-patterning, and other cutting-edge technologies, TSMC will be able to achieve higher transistor density, improved performance, and reduced power consumption in its semiconductor chips. TSMC's collaboration with leading semiconductor companies, such as Apple, AMD, and NVIDIA, highlights the importance of advanced lithography and process innovations in enabling next-generation electronic devices.

8.  **GLOBALFOUNDRIES**

- **Technology:** GLOBALFOUNDRIES is a leading semiconductor manufacturer offering advanced process technologies and manufacturing services [53].
- **Case study:** GLOBALFOUNDRIES' development of 12 nm FD-SOI (fully depleted silicon-on-insulator) process technology demonstrates its focus on delivering innovative solutions for power-efficient semiconductor devices. By utilizing FD-SOI technology, GLOBALFOUNDRIES [54] enables the fabrication of high-performance, low-power chips for applications such as IoT, automotive, and wireless connectivity. FD-SOI technology leverages advanced lithography techniques and process optimizations to achieve superior power efficiency and performance compared to traditional bulk CMOS processes [55].

9.  **Samsung Electronics**

- **Technology:** Samsung Electronics is a global leader in semiconductor manufacturing, offering a wide range of memory and logic products [56].

- **Case study:** Samsung's development of 3 nm GAAFET (gate-all-around FET) process technology represents a significant breakthrough in semiconductor fabrication. By adopting GAAFET transistor structures and advanced lithography methods, Samsung aims to deliver unprecedented levels of performance, energy efficiency, and scalability in its semiconductor devices. The integration of EUV lithography and innovative materials engineering enables Samsung to achieve finer feature sizes and higher transistor densities, paving the way for future generations of high-performance computing and mobile devices.

These are just a few examples of companies at the forefront of developing and deploying advanced technologies in semiconductor manufacturing. Their case studies highlight the importance of innovative lithography, metrology, process control, and materials solutions in advancing semiconductor technology and meeting the demands of next-generation electronic devices.

## 2.2 Overview of the Shift-left Approach

This book will explore the crucial theme of the semiconductor manufacturing process, possible defect scenarios with case studies, defect prevention in semiconductor manufacturing, emphasizing the transformative potential of AIAI, IoT, and emerging trends [20, 21, 22, 23, 26]. Through theoretical discussions, case studies, and practical examples, it has uncovered some trends which would be beneficial for semiconductor manufacturing processes, defect optimizing relating to that and integration with AI, ML, IoT. Let us investigate one by one.

The shift left approach may be based on proactively addressing defects early in the manufacturing process, facilitated by AIAI and IoT, and significantly improves yield [71] and reduces costs.

- **AI and IoT integration:** Data-driven insights from AI and real-time monitoring through IoT enable precise defect identification, prediction, and prevention.
- **Emerging trends:** Technologies like edge computing, fog computing, and smart manufacturing further enhance data processing and decision-making at the device level.
- **Case studies:** Practical examples demonstrate the effectiveness of AIAI, IoTIoT, and emerging trends in addressing specific defect scenarios across various processes.
- **Hidden costs of human error:** Proactive defect prevention strategies mitigate human error risks and associated costs, improving overall quality and reliability.

## 2.2.1 Implications for the future of semiconductor manufacturing

This book's findings have significant implications for the future of semiconductor manufacturing:

- **Increased yield [25, 71] and quality:** AIAI-powered defect prevention leads to higher yields, lower costs, and improved device reliability.
- **Enhanced process control:** Real-time monitoring and data-driven insights enable precise process control and optimization.
- **Shift to smart factories:** Integration of AIAI, IoT, and emerging technologies fosters smarter, data-driven manufacturing environments.
- **Improved sustainability:** Minimizing defects reduces waste and resource consumption, promoting sustainable manufacturing practices.
- **Evolving skillsets:** The workforce will need to adapt to new technologies and embrace data-driven decision-making.

# 3

# Overview of Semiconductor Manufacturing

Semiconductor manufacturing, also referred to as semiconductor fabrication or processing, is the intricate process of creating integrated circuits (ICs) or microchips that serve as the foundational components of electronic devices. This process entails a series of meticulously executed steps aimed at producing semiconductor devices with precise electronic properties and features.

## 3.1 Wafer Fabrication

Wafer preparation [1, 2] begins with preparing silicon wafers [1], which act as the substrate for semiconductor device fabrication. These wafers undergo various treatments to attain desired surface properties and cleanliness. We discuss this in the sections below.

### 3.1.1 Thin film deposition

Deposition [1, 2, 8] and patterning of metalmetal layers. Electrical isolation of metal layers in turn will require dielectric layers being deposited which can be done by the following methods shown in Figure 3.1. Let's check out an overview of all the terms we will be looking into detail in later sections.

Thin film deposition is a collection of techniques used to create a thin layer (coating) of material on a substrate surface. These films can range in thickness from a few nanometers to tens of micrometers and play a crucial role in various technological advancements.

Figure 3.1: Types in Thin Film Deposition.

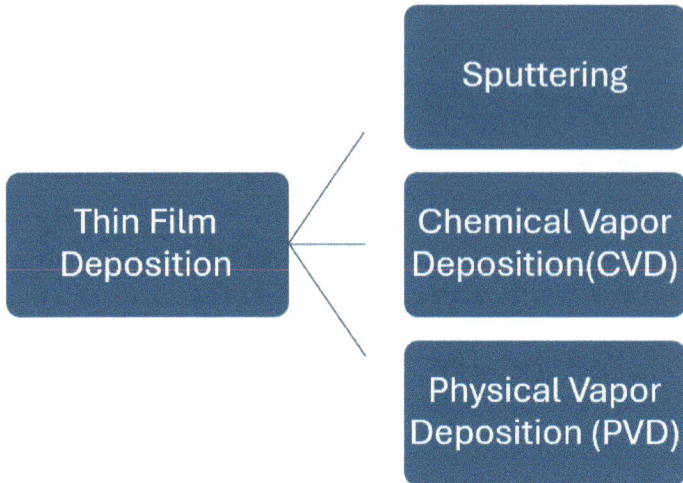

Thin film [8] deposition is the scientific art of creating and laying down extremely thin layers of material (usually ranging from a single atom to a few micrometers) onto a substrate surface. These coatings significantly alter the properties of the underlying material, enabling a wide range of applications in electronics, optics, and more. It involves depositing thin layers of materials like silicon dioxide and metal onto the wafer surface using techniques [2] such as chemical vapor deposition (CVD) and physical vapor deposition [2] (PVD).

**Evaporation:** In this method, the source material is heated until it vaporizes. The vaporized atoms travel through a vacuum chamber and condense onto the substrate, forming the thin film [1, 2, 8].

**Sputtering:** Here, a high-energy plasma bombards a target material, ejecting atoms that travel and deposit on the substrate (Figure 3.2). Different sputtering techniques like magnetron sputtering use magnets to confine the plasma for better control and deposition rate [1, 2, 8].

**Precursor reaction:** CVD involves introducing reactive gases into a vacuum chamber. These gases undergo chemical reactions near the substrate's surface, and the desired film material forms as a result [1, 2, 8]. See Figure 3.3.

**Figure 3.2:** Hypothetical image of high-energy plasma bombarding a target material.

**Figure 3.3:** Hypothetical figure of the precursor reaction.

### 3.1.1.1 Benefits of thin film deposition

Thin film deposition [57, 58, 59, 71, 72, 73, 74, 75, 76, 77] unlocks a treasure trove of advantages:

- **Tailored properties:** Deposited thin films can modify the substrate's conductivity, reflectivity, wear resistance, and other characteristics, enabling diverse functionalities.
- **Precise control:** Deposition techniques offer exceptional control over film thickness and composition, allowing for precise engineering of material properties.
- **Broad applications:** Thin films are ubiquitous in modern technology, from creating microchips and solar cells to applying anti-reflective coatings on lenses and wear-resistant layers on cutting tools.

### 3.1.1.2 Challenges to overcome: Considerations for thin film deposition

Despite its advantages, thin film deposition [72] comes with its own set of hurdles:

- **Vacuum requirement:** Most techniques necessitate a vacuum environment, adding complexity and cost to the process.
- **Material compatibility:** Compatibility between the substrate and deposited material is crucial to ensure proper adhesion and film integrity.
- **Process control:** Precise control over deposition parameters like temperature, pressure, and gas flow is essential for achieving desired film characteristics.

### 3.1.1.3 Basic mathematical equations used in different aspects of thin film deposition

This section deals with some forms of mathematical [72, 73, 74, 75, 76, 77] equations important to govern the critical dimensions in semiconductor manufacturing. However, one equation can be represented in multiple forms with different constants and variables (which can be found in later sections).

**1. Evaporation rate**

In evaporation, the rate at which the source material transitions to vapor is crucial. The Hertz–Knudsen equation describes this relationship:

$$F = \sqrt{\frac{P \cdot M}{2\pi kT}} \qquad (3.1)$$

where

- $F$: Evaporation rate (atoms/m$^2$ s)
- $P$: Vapor pressure of the source material (Pa)
- $M$: Molar mass of the source material (kg/mol)
- $K$: Boltzmann constant (1.38 $\times$ 10$^{-23}$ J/K)
- $T$: Temperature of the source material (K).

## 2. Sputtering yield

In sputtering, yield [71, 72] refers to the number of ejected atoms per incident ion. The following equation relates the yield ($Y$) to the sputter gas pressure ($P$), the target material's atomic mass ($M_t$), and the energy of the incident ions ($E_i$):

$$Y = Y_n \left(\frac{P}{P_0}\right)\left(\frac{E_i}{E_{th}}\right)^{n-1} \tag{3.2}$$

Here

- $Y_n$ : Reference sputtering yield at a specific pressure ($P_0$) and energy ($E_{th}$)
- $P_0$: Reference pressure
- $E_{th}$ : Threshold energy for sputtering the target material.

## 3. Film thickness in CVD

Chemical vapor deposition relies on the reaction rate of precursor gases to determine the film thickness. The following equation [72, 73, 74, 75, 76, 77] estimates the film thickness ($h$) based on the reaction rate ($R$), deposition time ($t$), and the density of the deposited film ($\rho$):

$$h = \frac{R \cdot t}{\rho}. \tag{3.3}$$

## 4. Thin film stress

Deposited thin films can experience stress due to factors like lattice mismatch with the substrate or thermal expansion differences. The Stoney [73] equation relates the film stress ($\sigma$) to the curvature ($\kappa$) of the substrate, the film thickness ($h$), and the Young's modulus of the substrate ($E_s$) and film ($E_f$):

$$\sigma = \frac{1}{6\,(1-\nu_s)} \cdot \left(\frac{E_f h^2}{1+\left(\frac{E_f}{E_s}\right)\left(\frac{h}{t_s}\right)}\right). \tag{3.4}$$

Here

$\nu_s$: Poisson's ratio of the substrate

$t_s$: Substrate thickness.

### 3.1.1.4 Application of these mathematical proportionalities in various scientific and engineering contexts

**Semiconductor manufacturing:** Thin film deposition is a cornerstone of creating integrated circuits (ICs) and microchips. The formulas can be used to:

- **Optimize the deposition [72, 73] rate:** Control the thickness and growth rate of thin film layers on transistors and other chip components. (Equation (3.1): Evaporation rate).
- **Tailor material properties:** Equations can help predict film stress (Equation (3.4)) resulting from material properties like Young's modulus ($E_f$ and $E_s$) to ensure film integrity and prevent device failures.

**Microelectromechanical systems (MEMS):** MEMS [74] devices like pressure sensors and accelerometers rely on precisely deposited thin films. The formulas can be used to:

- **Control film [71, 72, 73, 74] thickness:** Precise film thickness (Equation (3.3)) is crucial for MEMS device functionality. The equations can help determine deposition parameters to achieve desired thicknesses.
- **Manage stress:** Managing film stress (Equation (3.4)) is important in MEMSMEMS to maintain device performance and reliability.

**Optical coatings:** Thin films are used to create anti-reflective coatings for lenses, filters, and solar cells. The formulas can be used to:

- **Design reflectivity:** By understanding factors like film thickness and refractive index, engineers can design coatings with desired reflectance or transmittance properties [71, 72, 73, 74].
- **Optimize material choice:** Material properties like Young's modulus ($E_f$ and $E_s$) can be considered to ensure the deposited film adheres well to the substrate and withstands environmental stresses.

**Wear-resistant coatings:** Thin films can protect surfaces from wear and tear. The formulas can be used to:

- **Select the deposition technique:** Equations like those for evaporation rate (Equation (3.1)) and sputtering yield [71] (Equation (3.2)) can help choose deposition methods suitable for achieving the desired film density and wear resistance.
- **Control film properties:** Film thickness (Equation (3.3)) and stress (Equation (3.4)) can be crucial for wear resistance. The formulas can aid in optimizing deposition parameters for robust coatings.

**Best Usage:** These formulas are most effective when used in conjunction with material properties data and experimental results. Here are some best practices:

- **Identify key parameters:** Focus on the equation terms most relevant to your specific application (e.g., film thickness for MEMS, stress for wear-resistant coatings).
- **Material properties:** Obtain accurate material properties like Young's modulus [75] for the film and substrate to ensure reliable calculations.
- **Experimental validation:** Use the formulas as a starting point, but always validate results through experimental thin film deposition and characterization techniques.
- **Simulation tools:** Combine these formulas with simulation software to model thin film growth and predict film properties under various deposition conditions.

By understanding the underlying principles and best practices, these formulas become valuable tools for researchers, engineers, and scientists working in the field of thin film deposition.

### 3.1.1.5 Types of Thin Film Deposition Techniques: PVD

There are numerous thin film deposition techniques, each with its advantages and limitations.

The evaporation rate $(R_e)$ can be estimated using the Langmuir [76] equation:

$$R_e = P_s \left( \frac{A}{\sqrt{2\pi M R T}} \right) \exp \left( -\frac{E_v}{RT} \right) \qquad (3.5)$$

where

- $R_e$ : Evaporation rate (kg/m$^2$ s)
- $P_s$ : Vapor pressure of the material at temperature $T$ (Pa)
- $A$ : Surface area of the evaporating source (m$^2$)
- $M$ : Molar mass of the material (kg/mol)

- $R$: Universal gas constantgas constant (8.314 J/mol K)
- $T$: Temperature of the evaporating source (K)
- $E_v$ : Enthalpy of vaporization (J/mol):

Here's an overview of some of the most common methods:

## 1. Thermal evaporation

The target material is heated to a high temperature, causing it to evaporate (Figure 3.4) and deposit on the substrate. This technique is suitable for materials with high vapor pressures.

Thermal evaporation deposition encompasses heating a solid material within a high vacuum chamber until it reaches a temperature that generates vapor pressure. This vapor pressure, even at relatively low levels, forms a vapor cloud within the chamber. The evaporated material then condenses into a vapor stream, which travels through the chamber and adheres to the substrate as a coating or film.

Typically, in thermal evaporation processes, the material is heated to its melting point, rendering it liquid and usually situated at the chamber's bottom in an upright crucible. From this source, the vapor rises upward, while the substrates are positioned inverted at the chamber's top. This configuration ensures that the surfaces intended for coating face downward toward the heated source material, facilitating the deposition process.

To ensure proper film adhesion and control over various film properties, adjustments may be necessary. Fortunately, thermal evaporation system design and methodologies offer flexibility in manipulating several parameters. This allows process engineers to achieve desired outcomes regarding film thickness, uniformity, adhesion strength, stress, grain structure, as well as optical or electrical properties, among others.

## 2. Resistive evaporation deposition

One technique, commonly known as resistive evaporation deposition, utilizes a basic electrical resistive heating element or filament to evaporate the coating material. These resistive evaporation filaments come in various physical configurations, with many resembling "boats" made of thin sheet metal, typically composed of high-temperature metals like tungsten, featuring formed indentations or troughs where the material is loaded. The resistive filament heating source operates at low voltage for safety but requires very high current, often reaching several hundred amps.

**Figure 3.4:** Evaporation.

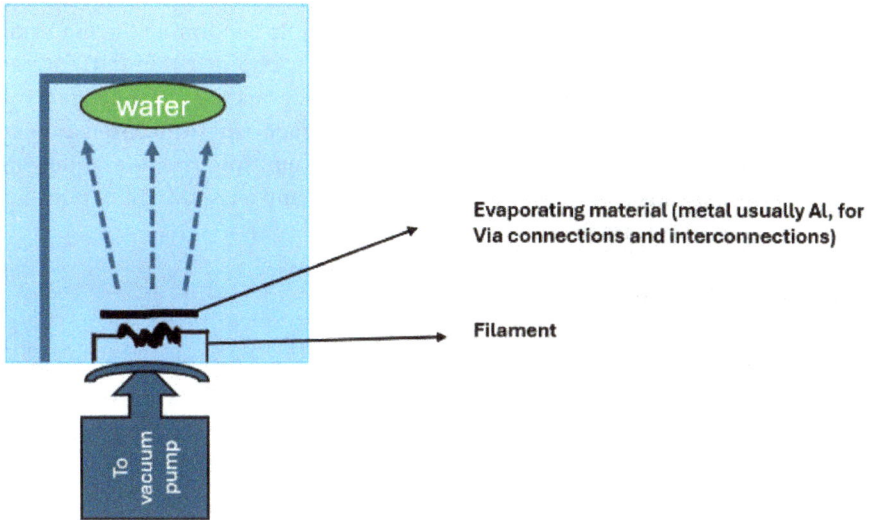

Resistive thermal deposition offers several advantages, including high deposition rates at a relatively low cost compared to other PVD processes. It presents a straightforward PVD coating method compatible with metals or non-metals/dielectrics such as chrome, aluminum, indium, gold, silver, calcium, lithium, and more. Additionally, it accommodates materials with low melting points and provides good directionality.

However, resistive thermal vapor deposition has its drawbacks. Film densities tend to be relatively low, although enhancements can be achieved with ion beam assisted deposition. While the equipment setup cost is comparatively lower, scalability is limited. Furthermore, compared to more complex PVD processes, it poses a higher risk of potential contamination.

### 3. E-beam

Another prevalent heat source used in vapor deposition is the electron beam, commonly referred to as e-beam evaporation. This method represents a more sophisticated approach to heating materials, involving high voltage (typically around 10,000 volts), necessitating additional safety precautions in e-beam systems. The primary component is the e-beam "gun," where a small and intensely hot filament emits electrons, which are then accelerated by the high voltage to form a high-energy electron beam.

This beam is magnetically guided into the crucible containing the material. At the standard 10 kV voltage, even a modest 0.1 amp of beam current can produce 1 kilowatt of concentrated power, effectively heating the material. The crucible is typically water-cooled to prevent damage from the intense heat. Commercially available e-beam guns often feature multiple crucibles, allowing for the simultaneous handling of several materials and easy switching between them for multi-layer processing.

### 4. Flash thermal evaporation

Flash thermal evaporation [71, 72, 73, 74, 75, 76] involves the utilization of a fine wire or powder made of coating material, which is often introduced into a heated ceramic crucible or heating element. Upon the application of power or upon contact with the hot element, the material evaporates almost instantaneously. When using crucibles, flash thermal evaporation equipment tends to offer faster deposition rates and thicknesses compared to methods employing wires. The manner in which coating powders are introduced into the crucible plays a pivotal role in ensuring the consistent thickness of films.

Since the material for the thin film is physically removed from a target in a vacuum chamber and deposited on the substrate, this is also referred to as physical vapor deposition/PVD.

PVD methods offer good control over film thickness and composition.

### 5. Sputtering

A high-energy beam (usually ionized gas) bombards the target material, ejecting atoms that deposit on the substrate. Sputtering allows for deposition of a wider range of materials compared to thermal evaporation [57, 58, 59, 71, 72, 73, 74, 75, 76].

Sputtering is a widely used physical vapor deposition (PVD) technique employed in the semiconductor, optics, and materials science industries for depositing thin films onto substrates. It involves the removal of atoms from a target material through the bombardment of energetic particles, typically ions or neutral atoms, onto the target surface[58].

The sputtering process occurs in a vacuum chamber filled with a low-pressure gas, usually argon. When high-energy ions from a plasma or cathode bombard the target material, they transfer momentum to the target atoms, dislodging them from the surface. These ejected atoms then travel through the vacuum and deposit onto the substrate, forming a thin film.

Mathematically, the sputtering yield ($Y$) can be expressed as:

$$Y = dN/dt = K \cdot I^n \tag{3.6}$$

where

- $dN/dt$ is the sputtering rate (number of atoms sputtered per unit time).
- $K$ is the sputtering constant, dependent on the target material and ion species.
- $I$ is the ion current density (ions per unit area per unit time).
- $n$ is the sputtering yield exponent, typically ranging from 0.5 to 2.

These equations represent simplified models and may require additional factors or adjustments for real-world applications. Simulation tools and experimental data play a crucial role in optimizing and predicting processes in microfabrication.

The sputtering rate is influenced by various parameters, including the ion energy, ion flux, target material properties, and chamber conditions. Higher ion energies and fluxes generally result in higher sputtering rates.

Sputtering is favored for its ability to deposit a wide range of materials, including metals, oxides, and nitrides, with excellent uniformity and purity. It is a versatile technique used in the manufacturing of semiconductor devices, solar cells, optical coatings, and magnetic storage media.

DC power may be used when depositing metals and RF supply is necessary when insulating films are deposited.

**Magnetron sputtering:** Magnetron sputtering has seen significant advancements in recent decades due to the growing demand for high-quality functional films in various industries. This technique has become a preferred choice for depositing coatings with superior performance compared to other methods [5, 7, 58, 59].

Initially developed in the 1970s, conventional magnetrons were a significant improvement over earlier sputtering processes but had limitations. The introduction of unbalanced magnetrons in the late 1980s and closed-field systems in the early 1990s revolutionized the industry by offering highly versatile and efficient deposition technologies. Pulsed magnetron sputtering [57, 58, 59] (PMS) further enhanced capabilities [58] by enabling the deposition of defect-free coatings, particularly insulating materials, at high rates without arcing.

Magnetron sputtering involves bombarding a target plate with energetic ions from plasma, causing the expulsion of target atoms which then condense onto a substrate as a thin film. Traditional sputtering methods suffer from low deposition rates and substrate heating issues, making them unsuitable for insulating materials [57].

In contrast, magnetron sputtering utilizes a magnetic field parallel to the target surface to trap secondary electrons near the target, increasing plasma density and leading to higher deposition rates. Unbalanced magnetrons, with a "leaky" magnetic field, extend the plasma towards the substrate, resulting in higher ion currents and low-energy ions ideal for coating quality.

Researchers like Windows and Savvides pioneered the study of unbalanced magnetron technology, demonstrating its ability to achieve ion fluxes up to ten times higher than conventional magnetrons, making it highly effective for producing dense and high-quality films.

**Closed-field unbalanced magnetron sputtering:** Closed-field unbalanced magnetron sputtering [8] is utilized for uniform coating of intricate components at acceptable rates, often employing multiple sources. In such systems, magnetic fields can be configured with either identical or opposite polarities. In the latter configuration, known as "closed-field," the linked magnetic fields create a high-density plasma region around the substrate, significantly enhancing deposition efficiency compared to mirrored configurations.

These multiple magnetron systems are well-suited for depositing multi-component materials, allowing each target to be composed of different materials and reactive gases to produce high-quality nitrides, oxides, etc. By adjusting sputtering rates or gas mixtures during deposition, alloy compositions and coating properties can be finely tuned, offering versatility in material design.

**Pulsed magnetron sputtering (PMS):** Pulsed magnetron sputtering (PMS) addresses challenges associated with depositing insulating materials, such as target charging, arcing, and stoichiometry control. By pulsing the magnetron discharge in the medium frequency range, arcs are minimized, improving deposit quality while achieving deposition rates comparable to pure metals. Despite the sophistication and cost of PMS power supplies, they have become the preferred choice for challenging insulating materials, producing dense, defect-free coatings with precise compositions.

### 6. Chemical vapor deposition

Chemical vapor deposition (CVD[8, 57, 58, 59, 71]) is a process used to produce high-quality thin films or coatings by chemical reactions in the vapor phase. It

**Figure 3.5:** Sputtering.

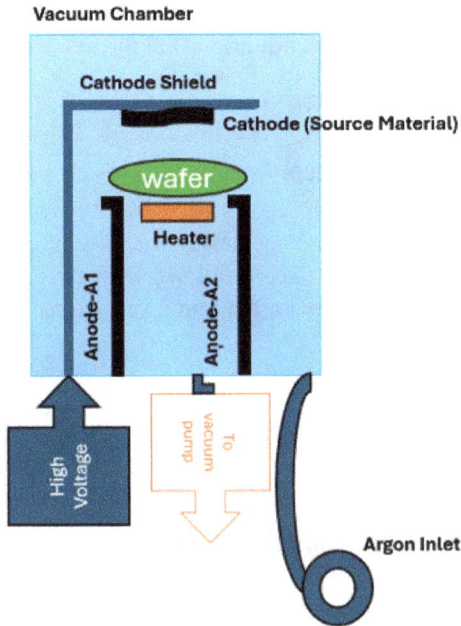

involves the deposition of solid material from a gaseous precursor onto a substrate surface. CVD offers precise control over film composition, thickness, and uniformity, making it widely used in various industries such as semiconductor manufacturing, optics, and materials science.

There are several types of CVD [77] techniques, each with its unique principles, advantages, and applications:

1. **Thermal CVD:** In thermal CVD, the precursor gases are heated to high temperatures (typically 500 °C to 1200 °C) to initiate chemical reactions leading to film deposition. The high temperatures provide the necessary energy for precursor molecules to decompose and react on the substrate surface. Thermal CVD is commonly used for depositing silicon dioxide (SiO2) and silicon nitride (Si3N4) films in semiconductor fabrication [71, 72, 73, 74, 75, 76].

    Mathematical equation [77] for thermal CVD:

    $$2\text{SiH}_4(\text{g}) + \text{O}_2(\text{g}) \rightarrow 2\text{SiO}_2(\text{s}) + 4\text{H}_2(\text{g}). \tag{3.7}$$

2. **Low-pressure CVD (LPCVD):** LPCVD operates at reduced pressures compared to thermal CVD, typically in the range of 0.1 to 10 Torr. Lowering the pressure allows for better control over the deposition process and results in films with improved uniformity and purity. LPCVD

is commonly used for depositing polysilicon, silicon dioxide, and silicon nitride films in semiconductor manufacturing.

Mathematical equation [77] for LPCVD:

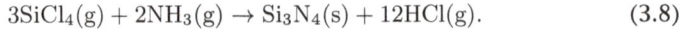

$$3\text{SiCl}_4(\text{g}) + 2\text{NH}_3(\text{g}) \rightarrow \text{Si}_3\text{N}_4(\text{s}) + 12\text{HCl}(\text{g}). \tag{3.8}$$

3. **Plasma-enhanced CVD (PECVD)**: PECVD utilizes plasma to enhance chemical reactions and reduce the required deposition temperatures. Plasma is generated by applying radiofrequency (RF) or microwave energy to the precursor gases, leading to the formation of highly reactive species that facilitate film deposition at lower temperatures (typically below 400 °C). PECVD is widely used for depositing thin films in microelectronics, optoelectronics, and photovoltaic devices.

Mathematical equation [77] for PECVD:

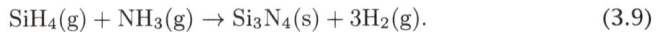

$$\text{SiH}_4(\text{g}) + \text{NH}_3(\text{g}) \rightarrow \text{Si}_3\text{N}_4(\text{s}) + 3\text{H}_2(\text{g}). \tag{3.9}$$

4. **Atomic layer deposition (ALD)**: ALD is a precise CVD technique that relies on self-limiting surface reactions to deposit ultra-thin films with atomic layer precision. ALD alternates between exposure to two or more gaseous precursors, allowing for controlled growth at the atomic level. ALD is commonly used for fabricating thin films in nanotechnology, MEMS (micro-electro-mechanical systems), and advanced materials research.

Mathematical equation for ALD [77]:

$$\text{Zn}(\text{CH}_3)_2(\text{g}) + \text{H}_2\text{O}(\text{g}) \rightarrow \text{ZnO}(\text{s}) + 2\text{CH}_4(\text{g}). \tag{3.10}$$

5. **Metalorganic CVD (MOCVD)**: MOCVD utilizes metalorganic precursors containing metal–carbon bonds which decompose at elevated temperatures to deposit thin films of metal oxides, nitrides, or alloys. MOCVD is extensively used for growing epitaxial layers of compound semiconductors such as gallium nitride (GaN) and indium phosphide (InP) for optoelectronic and LED (light-emitting diode) applications.

Mathematical equation [77] for MOCVD:

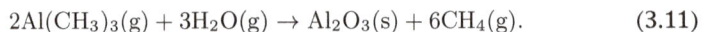

$$2\text{Al}(\text{CH}_3)_3(\text{g}) + 3\text{H}_2\text{O}(\text{g}) \rightarrow \text{Al}_2\text{O}_3(\text{s}) + 6\text{CH}_4(\text{g}). \tag{3.11}$$

Reaction rate [77] equation for CVD:

$$\text{Rate} = k_f [A]^m [B]^n \exp\left(-\frac{E_a}{R_g T}\right) \tag{3.12}$$

**Explanation of variables:**

- Rate: This represents the rate at which the reaction between the precursor chemicals (A and B) occurs on the substrate surface, leading to the formation of the desired film material. Units can vary depending on the specific reaction and product (e.g., $mol/cm^2/s$, $molecules/cm^2/s$).
- $k_f$ : This is the pre-exponential factor, a constant specific to the reaction and material system. It represents the frequency of successful collisions between reactant molecules at a given temperature.
- $[A]$ and $[B]$: These represent the concentrations of the precursor chemicals (A and B) at the reaction site on the substrate surface. Units are typically in moles per liter (mol/L). The exponents ($m$ and $n$) indicate the reaction order with respect to each precursor.
- $E_a$ : This is the activation energy, the minimum energy required for the reaction to proceed at a significant rate. Units are in Joules per mole (J/mol).
- $R_g$ : This is the universal gas constant, a constant value equal to 8.314 J/mol K.
- $T$: This is the absolute temperature of the reaction zone near the substrate surface, typically in Kelvin (K).

**Understanding the equation:**

This equation captures the dependence of the CVD reaction rate on several factors:

- **Precursor concentration:** Higher concentrations of reactants generally lead to a faster reaction rate.
- **Reaction order:** The exponents ($m$ and $n$) indicate the reaction order with respect to each precursor. A first-order dependence ($m$ or $n$ equal to 1) implies that a doubling of the concentration results in a doubling of the reaction rate.
- **Temperature:** The exponential term with activation energyenergy ($E_a$ ) signifies that the reaction rate increases exponentially with temperature. However, excessively high temperatures can lead to undesired side reactions or decomposition of precursor molecules.

**Important note:**

This equation is a simplified representation and might need modifications to account for specific CVD processes. Real-world CVD involves factors like mass transport of precursors, surface adsorption/desorption, and film growth mechanisms. Simulation tools and experimental data are often used to optimize CVD processes for specific materials and film properties.

### 3.1.2 Photolithography

Utilizes a photosensitive material called photoresist to create patterns on the wafer surface by selectively exposing it to ultraviolet [3] light through a photomask [7].

**The steps: A symphony of light and chemistry**

1. **Substrate preparation (120–200 °C):** The process begins with a silicon wafer, meticulously cleaned to remove any contaminants that might interfere with the subsequent steps. A pre-bake at moderate temperatures (around 120–200 °C) helps eliminate moisture from the wafer surface [10, 43, 47].
2. **Photoresist coating (spin coating):** A light-sensitive liquid called photoresist is uniformly applied to the wafer's surface using a spin coating technique. The wafer spins rapidly, creating a thin and even layer of photoresist.
3. **Soft bake (90–110 °C):** Another baking step, typically at 90–110 °C, removes residual solvents from the photoresist, solidifying it for the next stage.

**Figure 3.6:** Hypothetical image of soft baking of a wafer.

4. **Photomask alignment and exposure:** A high-resolution mask, containing the desired circuit pattern in a chrome layer on a glass substrate, is meticulously aligned with the photoresist-coated wafer. Powerful ultraviolet (UV) light is then shone through the mask, selectively exposing and altering the photoresist properties in the exposed areas.

5. **Post-exposure bake (PEB) (120–180 °C):** A final baking step (120–180 °C) further solid-ifies the photoresist pattern. The exposed regions undergo a chemical change, making them more soluble in the following development step.

**Figure 3.7:** Wafer under UV right after photo resist coating.

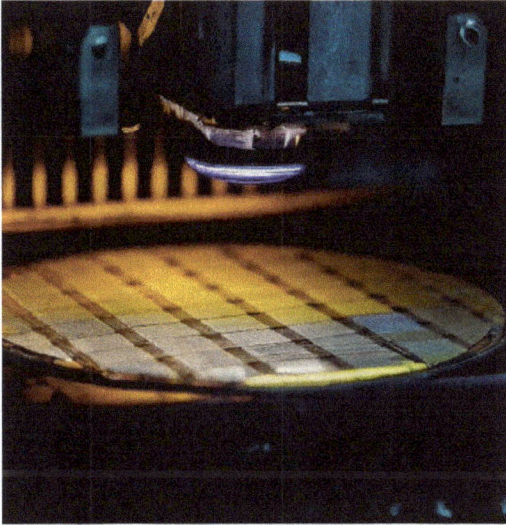

6. **Development:** The wafer is immersed in a developer solution, which selectively dissolves the exposed (typically more soluble) regions of the photoresist. This creates a patterned resist layer on the wafer, directly reflecting the circuit layout from the photomask.
7. **Pattern transfer (etching or deposition):** Now comes the magic! Depending on the desired outcome, either etching or deposition techniques are used to transfer the resist pattern onto the underlying silicon layer.
8. **Photoresist removal (strip):** Finally, the remaining photoresist is removed using a solvent, revealing the completed patterned layer on the silicon wafer.

### 3.1.2.1 The Science Behind the Light

Photolithography [10, 43, 47] leverages the photosensitivity of certain photore-sist materials. Upon exposure to UV light, the chemical structure of the exposed regions changes, altering their solubility in the developer solution. This allows for the creation of a patterned resist layer that serves as a blueprint for further processing.

**Temperature matters:** Throughout the process, specific temperatures play a crucial role:

- **Pre-bake and soft bake:** These steps remove solvents and solidify the photoresist, ensuring good adhesion and pattern fidelity.
- **Post-exposure bake:** This bake further hardens the resist pattern, preparing it for the development stage.

**Figure 3.8:** Picture of lab in fabrication facility with tools following 6S.

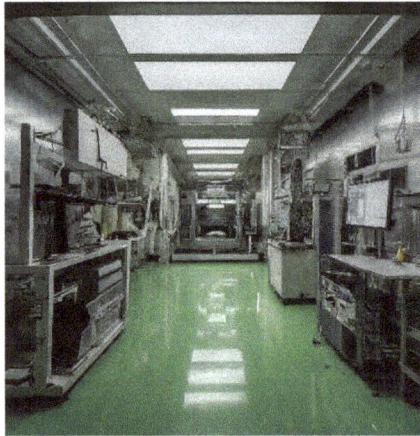

**Figure 3.9:** Photoetching technique - masking and exposure to UV radiation.

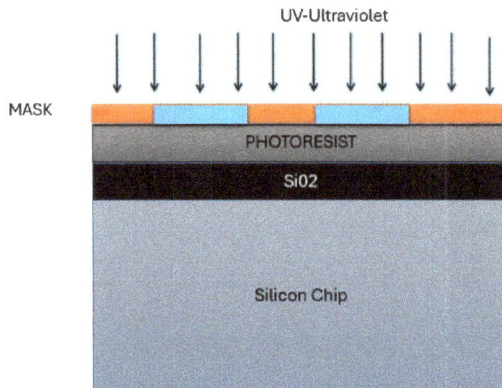

**Figure 3.10:** Photoetching technique – photoresist after development.

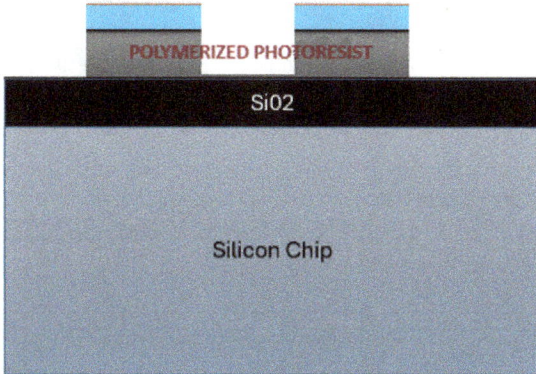

### 3.1.2.2 Mathematical equations used in different aspects of photolithography

While photolithography doesn't involve extensive mathematical modeling during the actual process steps, some equations can be used to understand and optimize certain aspects [10, 43, 47]:

1. **Film thickness in spin coating:** During spin coating (step 2), the photoresist thickness ($h$) on the wafer can be estimated using the following equation:

$$h = k\left(\nu \cdot t/\omega^2\right)^{1/3} \tag{3.13}$$

where

- $h$ – Thickness of the photoresist film (m)
- $k$ – Resist-specific constant (unitless)
- $\nu$ – Viscosity of the photoresist (Pa s)
- $t$ – Spin coating time (s)
- $\omega$ – Angular spin velocity (rad/s).

This equation helps determine the spin speed and time required to achieve a desired photoresist thickness for a specific resist material.

$$I = I_0 \cdot (1 - T) \cdot \left(\frac{\sin\left(\pi d/\lambda\right)}{\left(\pi d/\lambda\right)}\right) \tag{3.14}$$

where

- $I$ – Light intensity transmitted through the mask featurefeature (W/m$^2$)
- $I_0$ – Initial light intensity (W/m$^2$)
- $T$ – Mask opacity (0 for fully transparent, 1 for completely opaque)
- $d$ – Width of the mask feature (m)
- $\lambda$ – Wavelength of the UV light source (m).

2. **Photoresist dissolution rate:** The rate ($R$) at which the developer dissolves the exposed photoresist can be influenced by factors like the developer concentration ($C$) and temperature [10, 43, 47] ($T$). An empirical relationship might be used to model this:

$$R = k' \cdot C^n \cdot \exp\left(-\frac{E_a}{R_g T}\right) \qquad (3.15)$$

where

- $R$ – Dissolution rate (m/s)
- $k'$ – Pre-exponential factor (unit dependent on $n$)
- $C$ – Developer concentration (mol/L)
- $n$ – Empirical exponent
- $E_a$ – Activation energy for dissolution (J/mol)
- $R_g$ - Universal gas constant (8.314 J/mol K)
- $T$ – Developer temperature (K).

This equation (with appropriate fitting parameters) can help predict development times for achieving the desired pattern fidelity.

It's important to note that these are simplified examples, and actual photolithography [10, 43, 47] processes might involve more complex models and simulations considering factors like material properties, light diffraction effects, and resist chemistry.

### 3.1.3 Etching

Selectively removes exposed areas of the wafer using chemicalchemical or plasmaplasma etching processes, transferring the pattern onto the wafer surface [4, 9].

In the realm of microfabrication, etching plays a vital role in creating intricate structures on silicon wafers and other materials. It's akin to carving miniature features with incredible precision.

### 3.1.3.1 Wet etching (chemical etching)

- **Process:** A liquid etchant solution selectively dissolves specific materials on the wafer based on their chemical reactivity. This technique is often used for isotropic etching, where the etch [9] proceeds uniformly in all directions.

- **Examples:**

  - **Silicon dioxide ($SiO_2$) etching:** Hydrofluoric acid [42] (HF) is a common etchant for removing silicon dioxide layers used as masks or sacrificial layers during fabrication.
  - **Aluminum (Al) etching:** Wet etchants like phosphoric acid-based solutions are used to remove unwanted aluminum layers.

**Figure 3.11:** Process flow chart: etching.

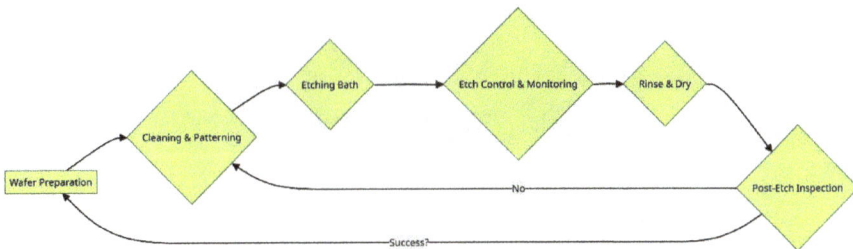

1. **Wafer preparation (A):** The silicon wafer is cleaned thoroughly to remove contaminants that might affect the etching process. A photoresist mask might be patterned onto the wafer to define the regions for etching.
2. **Cleaning and patterning (B):** Additional cleaning steps might be needed before etching, depending on the specific process.
3. **Etching bath (C):** The wafer is submerged in a suitable etchant solution that selectively removes the desired material.
4. **Etch controlControl and monitoring (D):** The etching process is carefully monitored (e.g., etch rate, time) to ensure it stops at the desired depth or feature profile.
5. **Rinse and dry (E):** The wafer is rinsed thoroughly to remove any residual etchant and then dried using techniques like nitrogen blow drying.

6. **Post-etch inspection (F):** The etched features are inspected using microscopy or other techniques to verify their dimensions and quality. If the results meet specifications, the process moves forward. Otherwise, adjustments might be needed in cleaning, patterning, or etch control parameters, leading back to step B.

**Figure 3.12:** Image of wafer under wet etch tank.

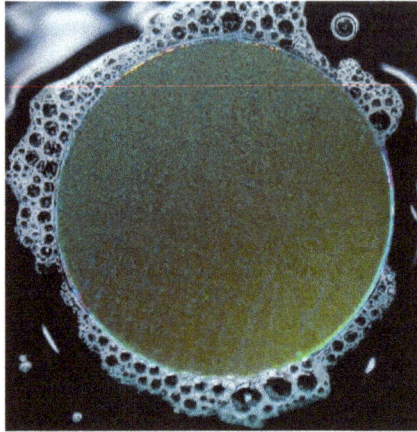

### 3.1.3.2 Dry etching (plasma etching)

**Process [4, 9, 42]:** This method utilizes plasma, a highly energetic gas containing charged particles (ions and electrons), to bombard the wafer surface. The ions physically sputter away material, enabling anisotropic etching (directional etching) with better control over feature profiles.

- **Examples:**

  - **Reactive ion etching (RIE):** A common dry etching technique where specific etching gases react with the wafer material, enhancing etch selectivity. For instance, chlorine-containing gases can be used for silicon etching.
  - **Deep reactive ion etching (DRIE):** A specialized RIE process that allows for etching deep trenches and high aspect ratio features crucial for microfluidic devices and MEMS (microelectromechanical systems).

**Figure 3.13:** Dry etch process steps.

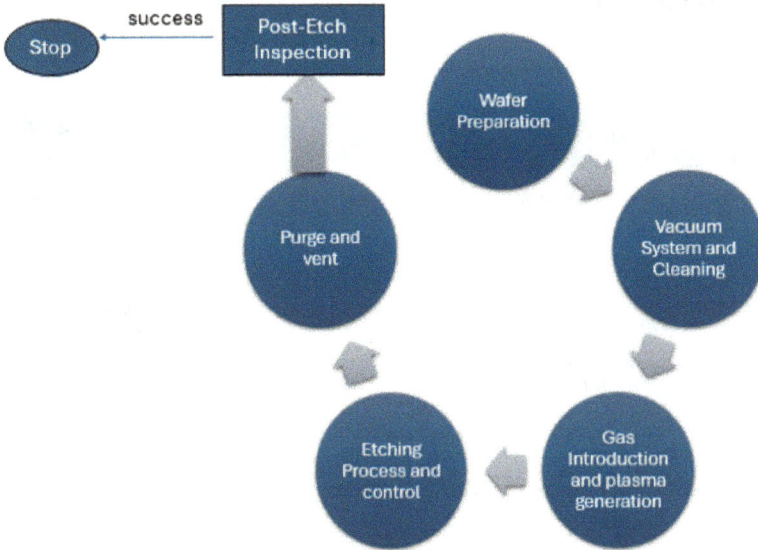

1.  **Wafer preparation (A):** Similar to wet etching, the wafer undergoes thorough cleaning.
2.  **Vacuum system and cleaning (B):** The wafer is loaded into a vacuum chamber, and a plasma cleaning step might be conducted to remove surface contaminants.
3.  **Gas introduction and plasma generation (C):** Specific etching gases are introduced into the chamber, and a high-frequency electric field is applied to create plasma.
4.  **Etching process and control (D):** The plasma bombards the wafer surface, selectively etching the exposed material. Parameters like etch time, power, and gas flow are carefully controlled.
5.  **Purge and vent (E):** After etching, the chamber is purged with an inert gas (e.g., nitrogen) and then vented to remove any remaining reactive species.
6.  **Post-etch inspection (F):** Similar to wet etching, the etched features are inspected to ensure they meet the desired specifications. Adjustments might be necessary if the results are not satisfactory.

### 3.1.3.3 Anisotropic vs. isotropic etching

**Anisotropic etching:** This technique [4, 9, 42] creates features with well-defined sidewalls, ideal for structures like trenches and pillars. Dry etching methods like RIE and DRIE offer superior control for anisotropic etching.

**Isotropic etching:** Wet etching often results in isotropic etching, where the material removal occurs in all directions at the same rate. This can be beneficial for creating rounded features or undercuts [78].

**Table 3.1:** Types of etching.

| Feature | Wet etching | Dry etching |
|---------|-------------|-------------|
| Process | Chemical reaction with liquid etchant | Physical sputtering by plasma ions |
| Selectivity | Moderate | High |
| Etching direction | Isotropic (mostly) | Anisotropic (better control) |
| Applications | Removing sacrificial layers, bulk micromachining | Etching trenches, vias, high aspect ratio features |
| Advantages | Simpler setup, lower cost (often) | High precision, better feature control |
| Disadvantages | Limited etch selectivity isotropic etching can be undesirable | Complex equipment, higher cost, potential for damage |

### 3.1.3.4 Ion-beam etching

**Figure 3.14:** Ion implantation basic block diagram for functionality.

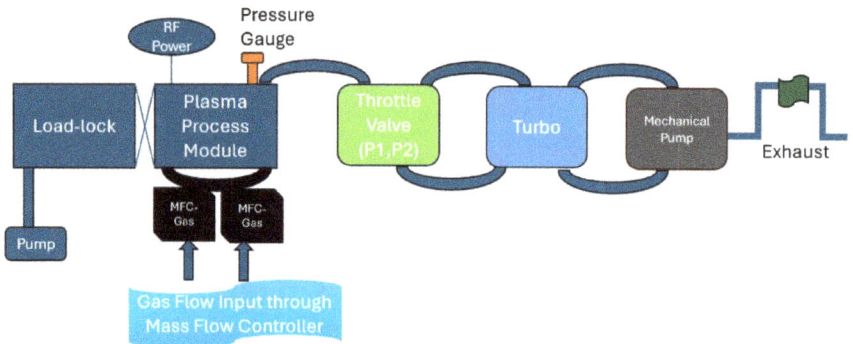

Plasma [42], often referred to as the fourth state of matter following solids, liquids, and gases, comprises a dynamic mixture of ions, electrons, and neutral species. Understanding the properties of these particles is crucial for harnessing their utility effectively.

- **Electrons:** These minute particles carry an electrical charge, allowing them to be swiftly accelerated using an electric field due to their low mass. This capability enables precise manipulation of electron trajectories.
- **Ions:** Considerably heavier than electrons, ions are generated when atoms or molecules are impacted by high-energy electrons. Their positive charge facilitates their control through electric fields, making them instrumental in processes like physical etching and metal deposition.
- **Radicals:** Formed through collisions with accelerated electrons, radicals are highly reactive species characterized by unpaired electrons. Their reactivity drives chemical etching reactions with materials, contributing to the overall etching process.

Etching involves a delicate balance between physical and chemical mechanisms, tailored to specific materials and desired outcomes.

Plasma etching terminology can be perplexing, often referred to as reactive ion etching (RIE) due to the involvement of reactive radicals and ions in the process.

Before exploring different configurations, understanding the prerequisites for plasma generation is essential. Plasmas are ubiquitous in our environment, from celestial bodies to artificial lighting, driven by various energy inputs. Specifically, plasmas generated by radio frequency (RF) power are of interest for applications like microwave ovens, televisions, and radar due to their advantageous properties.

Plasmas exist across a spectrum of pressure regimes, necessitating meticulous control over parameters like process pressure, gas flows, composition, and RF delivery for precise etching of three-dimensional structures. This control is achieved within a controlled vacuum chamber environment, incorporating vacuum pumps to maintain stability and facilitate efficient gas exhaust. Figure 3.14 illustrates a typical plasma processing system.

The diagram illustrates a load lock mechanism designed to segregate the chamber from external influences. While simpler systems lacking non-hazardous gas requirements are available without a load lock, most systems incorporate this feature. As automation levels increase, more advanced systems integrate sophisticated robots and multiple plasma process modules to enhance productivity and accommodate high-volume production.

Of particular interest in this context are the mechanisms governing ion and radical generation and their directed movement for specific tasks. Radio frequency (RF) serves as the primary energy source for plasma generation, with the choice of frequency dictated by regulatory bodies like the Federal Communication Commission (FCC) to avoid interference with essential communication, military, or medical bands. Frequencies commonly utilized range from

2 MHz to 13.56 MHz, although other frequencies are employed for specialized applications and desired plasma behaviors.

Early plasma etching systems employed RF sources connected across electrodes to form a capacitively coupled discharge plasma (CCP). Unlike in a DC discharge, distinguishing between anode and cathode in CCP is challenging. RF alternates high voltage, causing electron oscillation rather than a direct current, promoting ionization and dissociation of gases, resulting in higher plasma density compared to DC plasma. However, practical considerations like electrode size differences result in the smaller electrode acting as a cathode and the larger electrode typically grounded as the anode. A matching network is inserted between the RF source and the powered electrode to efficiently couple RF into the plasma through impedance matching and provide a blocking capacitor. This bias is crucial for directing ions and controlling their flux. Plasma ignition is induced by activating the RF, with conditions influenced by pressure, geometry, gas types, and electric fields generated.

In practice, wafers can be positioned on either the grounded or powered electrode, as depicted in Figure 3.15, showcasing the locations of the RF power supply and matching network. The gases, typically $CF_4$, $CHF_3$, $O_2$, $SF_6$, and $H_2$, can be used for dry etching. Plasma is generated with the help of an RF

**Figure 3.15:** Basic functionality showing dry etching through powered electrodes, generating ions.

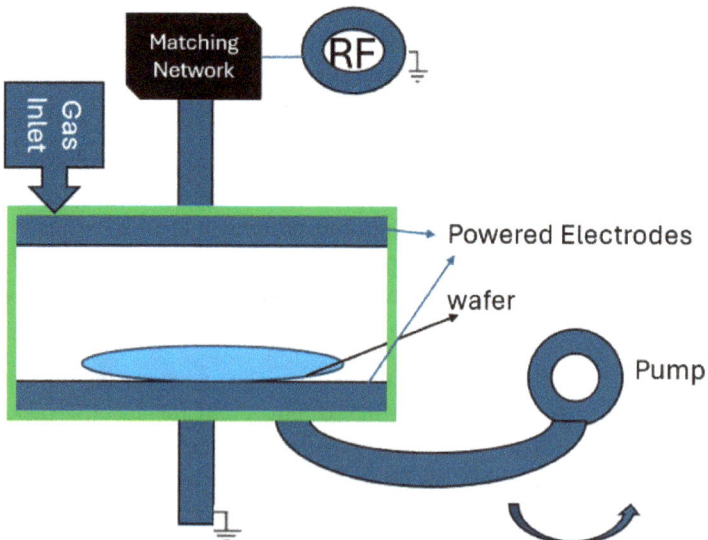

source. MFC, pressure (pump, throttle), and capacitive loads are the knobs that are used to control etching.

Critical dimensions and wafer profile, are some of the parameters that can be used to check the SPC details after the processes are completed.

### 3.1.3.5 Choosing the right technique

The selection of an etching technique depends on the desired feature profile, material properties, and process requirements. Wet etching might be suitable for simple structures or removing bulk material, while dry etching is preferred for high-precision, directional etching.

**The future of etching:** As microfabrication [79] pushes the boundaries of miniaturization, etching techniques are constantly evolving. Researchers are exploring novel methods like cryogenic etching (using extremely low temperatures) or using focused ion beams (FIB) for even more precise material removal [4, 9, 42, 78, 79].

### 3.1.3.6 Mathematical equations governing etching

1.  **(Wet) etch rate ($R$):** This equation describes the rate at which the material is removed by the etchant.

$$R = k \cdot C^n \cdot \exp\left(-\frac{E_a}{R_g T}\right) \tag{3.16}$$

where
$R$ – Etch rate ($\mu$m/min or nm/min)
$k$ – Pre-exponential factor (unit dependent on $n$)
$C$ – Etchant concentration (mol/L)
$n$ – Empirical exponent
$E_a$ – Activation energy for the etching reaction (J/mol)
$R_g$ – Universal gas constant (8.314 J/mol K)
$T$ – Etchant temperature (K).

Note: This is a simplified equation, and the actual etch rate can be influenced by factors like the material properties, etchant composition, and agitation during etching.

2.  **Selectivity ($S$):** This parameter indicates the relative etch rate between the desired material being etched and a mask material (or another material on the wafer).

$$S = \frac{R_{desired}}{R_{mask}} \tag{3.17}$$

where
$S$ – Selectivity (unitless)
$R_{desired}$ – Etch rate of the desired material (μm/min or nm/min)
$R_{mask}$ – Etch rate of the mask material (μm/min or nm/min).

A higher selectivity value indicates that the etchant preferentially removes the desired material with minimal attack on the mask layer.

3. **Sputtering yield ($Y$):** This parameter describes the number of atoms removed from the target material per incident ion in the plasma.

$$Y = f\left(E_i, \theta, M_t, M_i\right) \qquad (3.18)$$

where
$Y$ – Sputtering yield (atoms/ion)
$E_i$ – Energy of the incident ion (eV)
$\theta$ – Angle of incidence (degrees)
$M_t$ – Mass of the target material atom (kg)
$M_i$ – Mass of the incident ion atom (kg).

Sputtering yield depends on the ion energy, angle of incidence, and the masses of the target and incident ion species.

4. **(Dry) etch rate ($R$):** In dry etching, the etch rate can be expressed as:

$$R = Y \cdot \Phi \cdot \cos\left(\theta\right) \qquad (3.19)$$

where
$R$ – Etch rate (μm/min or nm/min)
$Y$ – Sputtering yield [71] (atoms/ion)
$\Phi$ – Ion flux (ions/cm$^2$/s)
$\theta$ – Angle of incidence (degrees).

The etch rate is a product of the sputtering yield, ion flux, and the cosine of the incident angle (accounting for the effective area exposed to the ions).

Important note: These are simplified mathematical models [79], and actual etching processes involve complex interactions between various parameters. Simulation tools and experimental data are often used to fine-tune etching processes for specific materials and feature profiles.

### 3.1.4 Doping/diffusion of impurities

Doping [1, 2, 5] is a fundamental step in semiconductor manufacturing, where carefully controlled amounts of impurities are introduced into a pure silicon crystal to alter its electrical conductivity. This strategic addition of impurities transforms the intrinsic (almost insulating) silicon into an extrinsic semiconductor with tailored electrical properties, forming the foundation of modern electronic devices. Here's a breakdown of the diffusion process for doping:

## 1. Wafer preparation (120–200 °C):

- The process begins with a silicon wafer, meticulously cleaned to remove any contaminants that might interfere with the subsequent diffusion step.
- A pre-bake at moderate temperatures (around 120–200 °C) helps eliminate moisture from the wafer surface, ensuring better diffusion uniformity.

## 2. Dopant source selection:

- The choice of dopant depends on the desired electrical properties.

  - Group III elements (boron, aluminum): These act as acceptor impurities, creating "holes" (positive charge carriers) and resulting in p-type (positive) semiconductors.
  - Group V elements (phosphorus, arsenic, antimony): These act as donor impurities, introducing additional electrons (negative charge carriers) and forming n-type (negative) semiconductors.

**Figure 3.16:** Basic diffusion process.

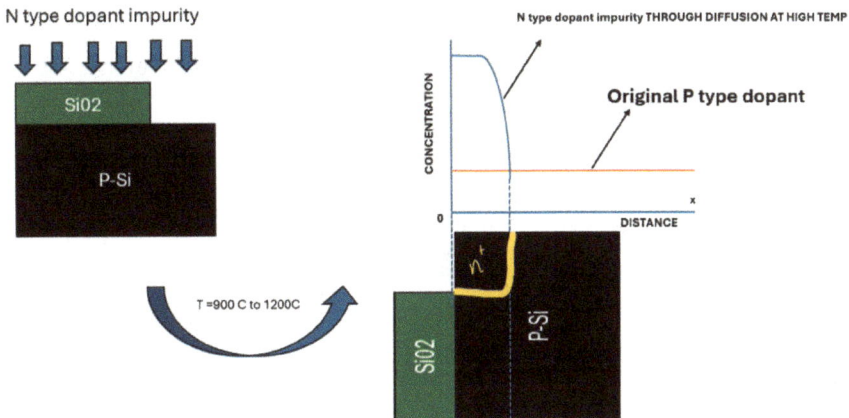

### 3. Diffusion [1, 2, 5, 79] techniques:

Several methods can be used to introduce dopant atoms into the silicon wafer:

- Solid-state diffusion:

  - The wafer is placed in a sealed container with a solid source material containing the dopant atoms.
  - The container is heated to a high temperature (typically 900–1200 °C) for a specific time.
  - At these elevated temperatures, dopant atoms from the solid source sublime (change directly from solid to gas) and diffuse into the silicon wafer.
  - The diffusion depth and concentration profile are controlled by factors like temperature, diffusion time, and dopant source concentration.

- Liquid-phase epitaxy (LPE):

  - A molten layer containing the dopant element is brought into contact with the silicon wafer.
  - Dopant atoms from the liquid phase diffuse into the underlying silicon.
  - LPE offers precise control over doping concentration but is less widely used compared to solid-state diffusion.

- Ion implantation:

  - This technique uses an accelerated beam of dopant ions to bombard the silicon wafer.
  - The ions penetrate the wafer surface and come to rest within the silicon lattice.
  - Ion implantation offers excellent control over dopant depth and profile but requires specialized equipment.

### 4. Post-diffusion processes:

- After the diffusion step, the wafer undergoes subsequent processes to activate the dopant atoms and create the desired device structures.
- These might involve additional heating steps (annealing) or etching procedures to define specific regions with different doping concentrations.

### The science behind diffusion [5]:

- Diffusion is a natural phenomenon where atoms or molecules move from a region of high concentration to a region of low concentration.

- In semiconductor doping, the dopant atoms diffuse from the source material into the silicon wafer due to a concentration gradient.
- Higher diffusion temperatures increase the atomic mobility and penetration depth of the dopant atoms.

### 3.1.4.1 Mathematical equations governing diffusion

**1. Fick's first law of diffusion:**

This fundamental equation [1, 2, 5, 79] describes the diffusive flux ($J$) of dopant atoms across a unit area per unit time:

$$J = -D^*(dc/dx) \tag{3.20}$$

where

- $J$ – Dopant flux ($cm^2/s$ or $m^2/s$)
- $D$ – Diffusion coefficient of the dopant in silicon ($cm^2/s$ or $m^2/s$) (depends on temperature and dopant species)
- $dc/dx$ – Concentration gradient of dopant atoms ($atoms/cm^3$ or $atoms/m^3$)
- $x$ – Distance from the wafer surface (cm or m).

The negative sign indicates that diffusion occurs from high concentration (positive $dc/dx$) to low concentration (negative $dc/dx$).

**2. Fick's second law of diffusion:**

This equation describes the time-dependent change in dopant concentration ($C$) within the silicon wafer:

$$dC/dt = D * d^2C/dx^2 \tag{3.21}$$

where

- $C$ – Dopant concentration at a specific location ($atoms/cm^3$ or $atoms/m^3$)
- $t$ – Time (s).

This equation is a partial differential equation and requires solving with appropriate boundary conditions to determine the dopant concentration profile within the wafer after a diffusion process.

**3. Solutions for simple diffusion cases:**

For specific diffusion scenarios with constant source concentration or a fixed total dopant amount, analytical solutions for the dopant concentration profile ($C(x, t)$) can be obtained. These solutions often involve error function (erf) or complementary error function (erfc) expressions.

Here are some examples of analytical solutions for simple diffusion cases involving the error function (erf) or complementary error function (erfc):

**[A] Constant source diffusion:**

Consider a scenario where the wafer surface is exposed to a constant dopant source concentration ($C_s$) for a diffusion time ($t$). This situation can be approximated by a semi-infinite solid model. The solution for the dopant concentration profile ($C(x, t)$) within the wafer is:

$$C\left(x, t\right) = C_s \left(1 - \mathrm{erf}\left(\frac{x}{2\sqrt{Dt}}\right)\right) \tag{3.22}$$

where

- ($C(x, t)$) – Dopant concentration at position $x$ and time $t$ (atoms/cm$^3$ or atoms/m$^3$)
- $C_0$ – Dopant concentration in the thin film source (atoms/cm or atoms/m$^3$)
- $x$ – Distance from the wafer surface (cm or m)
- $D$ – Diffusion coefficient of the dopant in silicon (cm$^2$/s or m$^2$/s)
- $t$ – Diffusion time (s).

**[B] Diffusion from a thin film source:**

If the dopant source is a thin film with a finite dopant concentration ($C_0$) deposited on the wafer surface, the solution for the dopant concentration profile ($C(x, t)$) can be expressed as

$$C\left(x, t\right) = \frac{C_0}{\sqrt{\pi Dt}} \exp\left(-\frac{x^2}{4Dt}\right) \tag{3.23}$$

where

- ($C(x, t)$) – Dopant concentration at position $x$ and time $t$ (atoms/cm$^3$ or atoms/m$^3$)
- $C_0$ – Dopant concentration in the thin film source (atoms/cm$^3$ or atoms/m$^3$)
- $x$ – Distance from the wafer surface (cm or m)
- $D$ – Diffusion coefficient of the dopant in silicon (cm$^2$/s or m$^2$/s)
- $t$ – Diffusion time (s).

**4. Semiconductor device simulation [1, 2, 5, 79] tools:**

Due to the complexities of real-world diffusion processes, advanced simulation tools are often used in semiconductor manufacturing. These tools numerically solve Fick's [80] second law, incorporating factors like temperature dependence of diffusion coefficients, multiple dopant species, and complex device geometries.

## 3.1.5 Oxidation

Semiconductor manufacturing relies on a series of intricate processes to create the tiny electronic components that power our modern world. One of the most fundamental steps in this process is *oxidation* [1, 2, 5, 79]. This note will delve into the significance of oxidation and its impact on semiconductor devices.

**What is oxidation?**

Oxidation refers to the process of forming a thin layer of oxide on the surface of a silicon wafer. In the context of semiconductors, this oxide is typically silicon dioxide ($SiO_2$), commonly known as silica. This layer is created by exposing the silicon wafer to an oxygen-rich environment at high temperatures.

**Why is oxidation important?**

The silicon dioxide layer formed during oxidation plays a critical role in various aspects of semiconductor device fabrication:

- **Insulation:** $SiO_2$ is an excellent insulator, preventing unwanted electrical current flow between different parts of the device. This is essential for the proper functioning of transistors and other electronic components.
- **Protection:** The oxide layer acts as a protective barrier, safeguarding the underlying silicon from contaminants that could degrade its electrical properties.
- **Foundation for further processing:** The $SiO_2$ layer serves as a base for subsequent steps in the fabrication process, such as photolithography and etching, which are used to define the intricate circuit patterns on the chip.

**Methods of oxidation:**

There are various methods for oxidizing a silicon wafer, with the most common being:

- **Thermal oxidation:** This technique involves exposing the wafer to high temperatures (typically 800–1200 °C) in an oxygen-rich environment, such as pure oxygen or water vapor.
- **Plasma [42]-enhanced chemical vapor deposition (PECVD):** This method utilizes a combination of plasma and chemical precursors to deposit the oxide layer at lower temperatures compared to thermal oxidation.

**Tailoring the oxidation Process:**

The thickness and properties of the oxide layer can be controlled by adjusting factors like oxidation time, temperature, and the specific oxidation method used. This allows for tailoring the oxide layer to meet the specific requirements of different devices.

Some other terms we commonly come across are:

- **Annealing:** Involves heat treatment processes to activate dopants, repair crystal defects, and relieve stress induced by previous processing steps [6].
- **Metallization:** Deposits metal layers onto the wafer surface to form interconnects and contact pads, facilitating connections between different components of the semiconductor device [11].
- **Dielectric layer deposition:** Deposition of insulating layers between metal layers to provide electrical isolation and prevent interference between different components [12].
- **Passivation:** Applies a protective layer to the wafer surface to shield it from contamination and damage during subsequent processing and packaging [13].

# 4

# Inline Parameters in Semiconductor Manufacturing

Inline parameters in semiconductor manufacturing, such as critical dimensions and thickness, play a crucial role in statistical process control (SPC) for monitoring and maintaining quality during wafer processing steps. These parameters are directly measured or inferred from various tools and techniques to ensure that the manufacturing process is within specified tolerances and yields consistent results. Here's how they contribute to SPC [19]:

## 4.1 Critical Dimensions (CD)

Critical dimensions refer to specific geometrical features on a semiconductor wafer that directly affect the functionality and performance of integrated circuits. These dimensions include line widths, spaces between lines, and various patterns etched or deposited on the wafer surface. Monitoring [19] CD variations through SPC helps maintain uniformity and precision in the fabrication process, ensuring that the final product meets design specifications.

In photolithography, the concept of critical dimensions (CD) refers to the minimum feature size that can be reliably transferred from the photomask onto the wafer. There isn't a single formula to determine the CD after processing, but several factors and parameters influence it. Here's a breakdown:

### 4.1.1 Factors affecting critical dimensions in lithography

- Wavelength (λ) of the light source: Shorter wavelengths allow for higher resolution and potentially smaller critical dimensions. Common light sources in photolithography include deep ultraviolet [3, 7, 10, 47, 43] (DUV) and extreme ultraviolet (EUV) light.
- Numerical aperture (NA) of the projection lens: A higher NA lens can focus light more tightly, leading to sharper feature definition and potentially smaller CDs.
- Photoresist properties: The resist thickness, contrast, and resolution capabilities play a crucial role. Thinner resists generally allow for smaller features, while high-contrast resists offer better pattern fidelity.
- Process parameters: Factors like baking temperatures, exposure times, and development conditions can all influence the final CD by affecting resist behavior.

**While there's no single formula for CD, some calculations can help estimate the minimum achievable feature size:**

**Rayleigh criterion:** This theoretical limit suggests the minimum feature size ($d_{min}$) for a given wavelength (λ) and numerical aperture (NA) based on the principles of diffraction:

$$d_{\min} \approx k^* \lambda / \text{NA} \tag{4.1}$$

where

- $d_{min}$ – Minimum feature size (m)
- $k$ – Factor related to the desired pattern profile (typically $k \approx 0.6$ for isolated lines)
- λ – Wavelength of the light source (m)
- NA – Numerical aperture of the projection lens (unitless).

The Rayleigh criterion is a theoretical limit, and actual achievable CDs might be larger due to practical limitations and resist behavior.

**Resolution enhancement techniques (RET):** These are advanced methods employed to push the boundaries of achievable feature sizes beyond the limits of the Rayleigh criterion. Some common RETs include:

- **Phase shifting masks:** Introduce phase shifts to enhance contrast and resolution.
- **Off-axis illumination:** Modulates the incident light to improve pattern fidelity.
- **Multiple patterning:** Creates features by combining multiple lithography steps.

**Measuring critical dimensions:**

After processing, critical dimensions are measured using specialized metrology tools like scanning electron microscopes (SEMs) or atomic force microscopes (AFMs). These tools provide high-resolution images and precise measurements of the feature sizes on the wafer.

In the context of etching, there isn't a single, universal formula to directly calculate the final dimensions (width and thickness) of a feature after etching. However, several factors and equations can help us understand and potentially predict the etch profile and resulting dimensions.

## 4.1.2 Factors affecting etch profile and dimensions

- **Etch type:** Wet etching typically exhibits isotropic etching (etches in all directions), while dry etching allows for anisotropic etching (directional etching). The chosen technique significantly impacts the final dimensions [4, 9, 42, 78].
- **Etch selectivity:** The ratio of etch rates between the desired material and the mask material influences how deeply the etch penetrates and the sidewall profile [4, 9, 42, 78].
- **Etch parameters:** Factors like etch time, power (for dry etching), etchant concentration (for wet etching), and temperature all affect the etch rate and depth.
- **Mask profile:** The initial profile of the photoresist pattern (lines, spaces, etc.) on the wafer acts as a template for the etching process.

**Figure 4.1:** Different die locations for wafer profile, critical dimensions measurements, inline params etc.

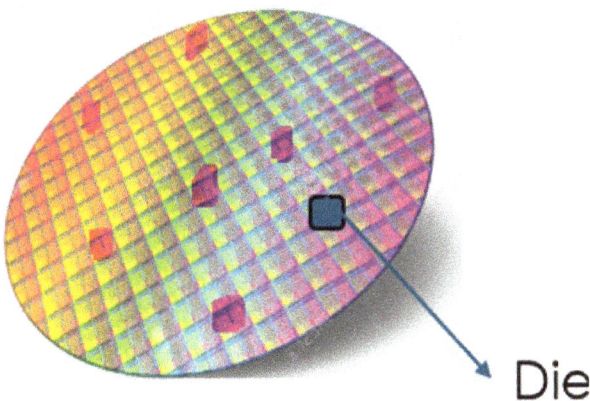

Different locations can be sampled to collect CD measurement instead of all dies inside the wafer (say 15% , 12% of the locations at different coordinates). These are fixed locations for the whole batch, so that there is no-location to location variation observed amongst batches. From these locations, a wafer profile is generated. The wafer profile can be compared against different wafers to check if there is uniformity. A non-uniform wafer profile can be indicative of defects, equipment parameters mismatched. Thus, this indicates a future possibility of these die being defective and causing issues after they are packaged as IC [15, 16].

**Figure 4.2:** Overall process mapping from wafer to chip.

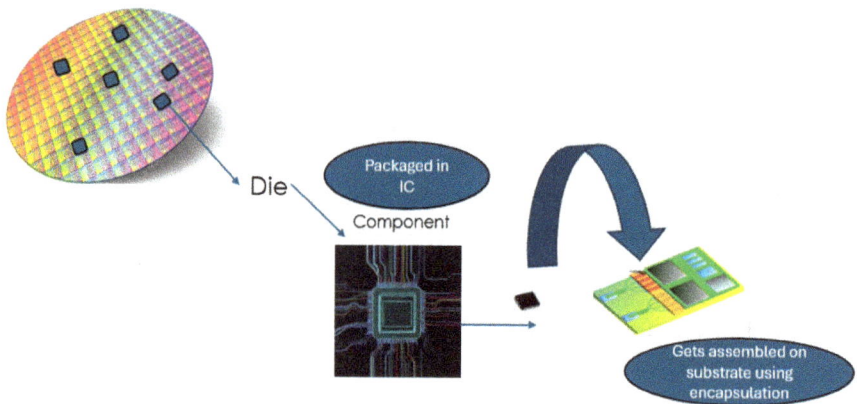

### 4.1.3 Diffusion parameters and dimensions critical to wafer functionality

1. **Diffusion [1, 2, 5, 79] length**

   - **Definition:** Diffusion length refers to the distance a dopant atom travels into the semiconductor material during the diffusion process. It indicates how deeply the dopant atoms penetrate the material.
   - **Importance:** The diffusion length determines the depth of the doped regions in the semiconductor, which directly affects device characteristics such as junction depth, threshold voltage, and breakdown voltage.
   - **Factors affecting:** Diffusion length is influenced by parameters such as the diffusion time, temperature, and dopant concentration. Longer diffusion times or higher temperatures can result in greater diffusion lengths.

2. **Diffusion time**

- **Definition:** Diffusion [1, 2, 5, 79] time is the duration for which the dopant atoms are allowed to diffuse into the semiconductor material. It determines the extent of dopant penetration into the semiconductor.
- **Importance:** The diffusion time directly impacts the depth of the doped regions and the concentration profile within the semiconductor material. Longer diffusion times allow for deeper dopant penetration, resulting in thicker doped layers.
- **Optimization:** Diffusion time needs to be carefully controlled to achieve the desired dopant profile. It is optimized based on the required junction depth and doping concentration for specific device applications.

3. **Diffusion temperature**

- **Definition:** Diffusion [1, 2, 5, 79] temperature refers to the temperature at which the diffusion process occurs. It plays a critical role in controlling the rate of dopant diffusion into semiconductor material.
- **Importance:** The diffusion temperature influences the diffusion rate and, consequently, the depth and concentration profile of the doped regions. Higher temperatures generally result in faster diffusion rates.
- **Temperature profile:** Diffusion processes often involve ramping up the temperature to facilitate dopant diffusion, followed by annealing steps to stabilize the dopant distribution and remove defects induced by diffusion.

4. **Diffusion concentration:**

- **Definition:** Diffusion [1, 2, 5, 79] concentration refers to the concentration of dopant atoms introduced into the semiconductor material during the diffusion process. It determines the doping level within the semiconductor.
- **Importance:** The diffusion concentration directly affects the electrical properties of the semiconductor, such as carrier concentration, conductivity, and junction characteristics. Higher dopant concentrations result in more heavily doped regions.
- **Uniformity:** Achieving uniform diffusion concentration across the semiconductor substrate is essential for ensuring consistent device performance and reliability. Precise control of the dopant source and diffusion process parameters is required to maintain uniformity.

### 4.1.3.1 Factors affecting these parameters

1. **Diffusion length**

   - **Temperature:** Higher temperatures generally result in greater diffusion lengths due to increased atomic mobility. For example, in a silicon wafer diffusion process, raising the temperature accelerates the movement of dopant atoms into the silicon lattice, leading to deeper diffusion profiles.
   - **Dopant species:** Different dopant species exhibit varying diffusion lengths in semiconductor materials. For instance, phosphorus atoms diffuse more readily into silicon compared to boron atoms due to their different atomic sizes and diffusion mechanisms.

2. **Diffusion time**

   - **Temperature:** Longer diffusion times are typically required at lower temperatures to achieve the desired dopant penetration depth. For example, in the fabrication of bipolar junction transistors (BJTs), diffusion times ranging from several hours to days may be necessary to achieve precise emitter-base junction depths at lower temperatures.
   - **Diffusion rate:** The rate of dopant diffusion increases with time, leading to deeper diffusion profiles. For instance, during the fabrication of shallow junctions in complementary metal-oxide-semiconductor (CMOS) technology, shorter diffusion times are employed to limit the dopant penetration depth and maintain junction integrity.

3. **Diffusion temperature**

   - **Time–temperature relationship:** Diffusion processes adhere to the Arrhenius equation, which describes the exponential relationship between diffusion rate and temperature. As temperature increases, the diffusion rate accelerates exponentially, leading to faster dopant movement and greater diffusion lengths. For example, in the formation of source/drain regions in MOSFET devices, elevated temperatures of around 900–1100 °C are typically employed to achieve sufficient dopant activation and diffusion.
   - **Dopant solubility:** The solubility of dopant atoms in the semiconductor matrix varies with temperature. At higher temperatures, dopant atoms have higher solubility, facilitating their incorporation into the semiconductor lattice. Conversely, at lower temperatures, dopant solubility decreases, leading to limited diffusion. For instance, in the fabrication of shallow junctions in advanced CMOS technology, ultra-low temperatures are utilized to suppress dopant diffusion and maintain precise junction depths.

4. **Diffusion concentration**

- **Dopant [5] source concentration:** The concentration of dopant atoms in the diffusion source material directly impacts the resulting diffusion concentration in the semiconductor substrate. Higher dopant source concentrations lead to higher diffusion concentrations in the semiconductor. For example, in the fabrication of heavily doped source/drain regions in MOSFET [12] devices, dopant sources with high concentrations of arsenic or phosphorus are utilized to achieve the desired doping levels.
- **Diffusion time and temperature:** Diffusion concentration is influenced by both diffusion time and temperature. Longer diffusion times and higher temperatures result in greater dopant incorporation into the semiconductor lattice, leading to higher diffusion concentrations. For example, during the formation of base regions in bipolar junction transistors (BJTs), extended diffusion times at elevated temperatures are employed to achieve the required doping levels for optimal device performance.

## 4.1.4 Important params

These parameters are also known as params, or inline params in semiconductor manufacturing terminology [12, 46].

**Thickness:** Thickness measurements are essential for layers deposited or etched during wafer processing, such as thin films, dielectrics, and metal layers. Variations in layer thickness can impact device performance, electrical characteristics, and reliability. By monitoring thickness parameters using SPC, manufacturers can detect process deviations, equipment malfunctions, or material variations early, allowing timely adjustments to maintain consistent layer thickness across wafers and lots.

**Uniformity:** Apart from CD and thickness, SPC also tracks the uniformity of critical parameters across the wafer surface. Uniformity metrics assess variations in CD, thickness, and other properties within a wafer (intra-wafer uniformity) and between different wafers in a lot (inter-wafer uniformity). Maintaining uniformity ensures that devices fabricated across the wafer or from different lots exhibit consistent performance and reliability, reducing defects and yield loss.

**Defect density:** Inline SPC also monitors defect density, which refers to the number of defects or anomalies present on the wafer surface per unit area. Defects can arise from process variations, material impurities, or equipment issues and can adversely affect device functionality and yield [46]. By analyzing defect density trends using SPC techniques, manufacturers can identify potential sources of defects [1, 2, 5, 17, 19, 46, 79], implement corrective actions, and improve overall process control and yield.

# 5

# Electrical Testing in Wafer, Die

Semiconductor devices are marvels of miniaturization, with features often smaller than the width of a human hair. Understanding and optimizing their behavior requires specialized techniques to probe their electrical characteristics, material properties, and physical dimensions. This note explores various probing methods, highlighting their differences, types, and the role of mathematical equations in interpreting the results [14, 46].

**Why probing [12, 19, 66, 71] is crucial:**

- **Device performance:** Probing helps evaluate a device's electrical parameters (current, voltage, resistance, capacitance) to ensure it meets performance specifications.
- **Material characterization:** Probing techniques can reveal material properties like doping concentration, carrier mobility, and defect densities, crucial for device optimization.
- **Troubleshooting and failure analysis:** Probing allows engineers to isolate faulty regions within a device and diagnose the root cause of malfunctioning.

## 5.1 Types of Probing Techniques

1. **Electrical probing:**

    - **Focus:** Measures electrical parameters like current, voltage, and resistance using specialized probes that make contact with specific points on the device.
    - **EquationsEquations:** Ohm's Law ($V = IR$) and Kirchhoff's current law ($\Sigma I_{in} = \Sigma I_{out}$) form the foundation for interpreting voltage and current measurements.

2. **Optical probing:**

- **Focus:** Utilizes various light-based techniques to probe devices. Examples include photoluminescence (PL) spectroscopy to analyze bandgap energies and electroluminescence (EL) imaging to visualize light emission patterns.

3. **Microscopy-based probing:**

- **Focus:** Employs high-resolution microscopes like scanning electron microscope (SEM) and transmission electron microscope (TEM) to examine the physical structure and morphology of devices at the nanoscale [16].

## 5.2 Mathematical Equations and Probing

While specific [12, 19, 66, 71, 79] equations depend on the chosen probing technique, here are some general examples:

- **Electrical probing:** Ohm's law ($V = IR$) is used to calculate resistance from voltage and current measurements.
- **Optical probing:** In PL spectroscopy, the peak energy of the emitted light can be related to the bandgap energy of the semiconductor material using Planck's equation ($E = hc/\lambda$).
- **TE (thermal engineering):** Heat transfer: The fundamental equation for heat transfer (Fourier's Law) relates heat flux ($q$) to temperature gradient ($dT/dx$):

$$q = -k^*(dT/dx) \tag{5.1}$$

where $k$ is the thermal conductivity of the material.

- **Microscopy-based probing:** Image analysis software might utilize geometric equations to quantify feature sizes based on pixel measurements in micrographs.

  - Resolution: Resolution in SEM is often described by the beam diameter and related to the accelerating voltage ($V$) and working distance ($WD$) through a complex function. However, a simplified approximation is:

$$\text{Resolution} \approx 0.05^* WD^*(V^{0.5}). \tag{5.2}$$

- **DC probingProbing:** In basic DC [81] probing, the measured voltage ($V_{measured}$) is the difference between the actual voltage ($V_{point}$) and the reference voltage ($V_{ref}$):

$$V_{\text{measured}} = V_{\text{point}} - V_{\text{ref}} \tag{5.3}$$

- **Laser voltage probing (LVP):** LVP uses a laser beam to induce a small voltage ($\Delta V$) proportional to the applied laser power ($P$) and a coefficient ($\alpha$):

$$\Delta V = \alpha^* P \qquad (5.4)$$

Table 5.1 shows the key differences between probing techniques [81].

**Table 5.1:** Key differences between probing techniques.

| Feature | Electrical probing | Optical probing | Microscopy-based probing |
|---|---|---|---|
| Focus | Electrical parameters | Material properties through light interaction | Physical structure and morphology |
| Measurement type | Quantitative (electrical signals) | Qualitative (spectra or images) | Qualitative (images) |
| Spatial resolution | Limited by probe size | Can be diffraction-limited | High resolution (nanoscale) |

## 5.3 Introduction to Testing and Packaging

- **Electrical testing:** Involves rigorous testing of each semiconductor device to ensure proper functionality and performance, including parametric, functional, and reliability tests [14, 81].
- **Die singulation:** Cuts the wafer into individual semiconductor devices or dies using techniques like sawing or laser cutting [15, 81].
- **Die attachment:** Mounts semiconductor dies onto substrates using adhesive materials or soldering techniques [16, 81].
- **Wire bonding:** Bonds thin wires between the semiconductor die and the substrate to establish electrical connections [17].
- **Encapsulation:** Encases the semiconductor device in a protective package, shielding it from environmental factors like moisture, dust, and mechanical shock [18].
- **Final testing:** Conducts comprehensive testing on packaged semiconductor devices to verify functionality, performance, and reliability before shipment to customers[19].
- In conclusion, semiconductor manufacturing is a sophisticated process requiring advanced technologies, expertise, and meticulous quality control measures to produce high-performance semiconductor devices used across various electronic applications [19, 81].

# 6

## Introduction to Process Optimization in Semiconductor Manufacturing Processes

Semiconductor manufacturing involves a complex series of steps, each suscep-tible to introducing defects that can affect device performance or yield [46, 71] (percentage of good devices per wafer). Here's an overview of common defects associated with different processes:

1. **Wafer preparation:**

   - **Contamination [45]:** Dust particles, organic residues, or metallic impurities can adhere to the wafer surface, causing leakage currents or shorts in devices.
   - **Surface scratches:** Scratches caused by improper handling or cleaning can introduce weak points in the oxide layer grown later.

2. **Oxidation:**

   - **Uneven oxide thickness:** Variations in oxide thickness can lead to non-uniform device characteristics and potential reliability issues [35, 36, 37].
   - **Interface defects [38]:** Imperfections at the silicon–oxide interface can trap charge carriers, affecting device performance.

3. **Lithography:**

   - **Photoresist patterning issues:** Underexposure or overexposure of photoresist during photolithography can result in undesired circuit patterns, affecting device functionality.
   - **Mask defects:** Imperfections on the photolithography mask can be transferred to the wafer, leading to unwanted circuit features.

4.  **Etching:**

    - **Over etching:** Excessive etching can remove desired features or lead to under-cuts, affecting device integrity.
    - **Anisotropic vs. isotropic etching issues:** Anisotropic etching (etches in one direction preferentially) might not create desired sidewall profiles, while isotropic etching (etches uniformly in all directions) might lead to pattern shrinkage.

5.  **Doping:**

    - **Non-uniform doping profile:** Uneven distribution of dopant atoms can lead to variations in device characteristics across the wafer [5].
    - **Doping level deviations:** Doping concentration that deviates from desired levels can significantly alter device behavior.

6.  **Metallization:**

    - **Voids or poor adhesion:** Gaps or weak bonding between metal layers can cause electrical discontinuities or reliability issues [11, 13].
    - **Metal spikes or protrusions:** Sharp features on metal layers can cause shorts or leakage currents.

7.  **Chemical mechanical polishing (CMP):**

    - **Non-uniform planarization:** Uneven removal of material during CMP can lead to height variations across the wafer, impacting device performance.
    - **Scratches or defects:** Scratches introduced during CMP can damage underlying layers or create leakage paths [40].

**Detection and Prevention:**

- **Inspection techniques:** Optical microscopy, electrical testing, and other methods are used to identify defects at various stages of processing.
- **Process control:** Statistical process control (SPC) techniques help monitor and optimize processing parameters to minimize defects [26].

*There is a different section dedicated to discussing semiconductor manufacturing defects later. This chapter intends to discuss in brief the overview of defects.*

## 6.1 Importance of Defect Prevention

Defect prevention is paramount in semiconductor manufacturing due to several critical reasons:

1. **Yield enhancement:** Defects in semiconductor devices can significantly reduce production yields, leading to increased costs and lower profitability. By preventing defects early in the manufacturing process, manufacturers can improve yield rates and maximize production efficiency.
2. **Quality assurance:** Defect-free semiconductor devices are essential for ensuring product quality and reliability. Even minor defects can compromise device performance and lead to product failures, potentially damaging a manufacturer's reputation and credibility in the market.
3. **Cost reduction:** Repairing defects in semiconductor devices after production is costly and time-consuming. Implementing effective defect prevention measures helps minimize the need for rework and scrap, resulting in cost savings for manufacturers.
4. **Time-to-market:** Timely delivery of high-quality semiconductor products is critical in meeting market demand and gaining a competitive edge. Defect prevention measures help streamline the manufacturing process, reducing cycle times and accelerating time-to-market for new products.
5. **Customer satisfaction:** Customers expect reliable and defect-free semiconductor devices that meet their performance requirements. Defect prevention ensures that customers receive high-quality products that perform as expected, leading to increased satisfaction and loyalty.
6. **Compliance with standards:** Semiconductor manufacturers must adhere to stringent quality standards and regulations imposed by industry bodies and regulatory authorities. Implementing defect prevention measures helps ensure compliance with these standards, avoiding penalties and legal implications.
7. **Risk mitigation:** Defects in semiconductor devices can pose significant risks to end-users, particularly in safety-critical applications such as automotive, aerospace, and medical devices. Preventing defects helps mitigate these risks and ensures the safety and reliability of electronic systems and devices.

## 6.2 Shift Left in Semiconductor Manufacturing

Shift left [1, 2, 25] refers to the concept of detecting and addressing issues as early as possible in the semiconductor manufacturing process. This approach involves moving quality assurance activities, such as testing and validation, to earlier stages of the manufacturing lifecycle, rather than waiting until the end.

The shift left philosophy recognizes that addressing defects and quality issues early in the process is more efficient and cost-effective than dealing with them later. By identifying and resolving issues during the design, development, and fabrication stages, manufacturers can prevent defects from propagating downstream and minimize their impact on production yields and product quality.

Key aspects of shift left in semiconductor manufacturing include:

1. **Early design verification:** Conducting comprehensive design verification and simulation studies to identify potential issues and optimize designs for manufacturability before fabrication begins.
2. **Process simulation and modeling:** Using advanced process simulation and modeling techniques to predict and optimize manufacturing processes, ensuring consistent and reliable production outcomes.
3. **Inline inspection and monitoring:** Implementing inline inspection and monitoring systems to detect defects and deviations from quality standards during the fabrication process, allowing for immediate corrective actions.
4. **Design for manufacturing (DFM):** Designing semiconductor devices with manufacturing considerations in mind, such as process variability, yield optimization, and defect tolerance, to minimize the likelihood of defects during fabrication [25].
5. **Continuous improvement:** Establishing a culture of continuous improvement and quality assurance throughout the semiconductor manufacturing lifecycle, with a focus on proactive defect prevention and process optimization [11].
   Overall, the shift left approach in semiconductor manufacturing [25] aims to enhance quality, reduce costs, and accelerate time-to-market by proactively addressing defects and quality issues early in the production process. By embracing this philosophy, semiconductor manufacturers can improve their competitiveness and deliver high-quality products that meet customer expectations.

## 6.3 Process Optimization in Semiconductor Manufacturing

Process optimization is a crucial aspect of semiconductor manufacturing that focuses on enhancing the efficiency, yield, and quality of semiconductor fabrication processes. It involves the continuous refinement and improvement of various manufacturing processes, equipment, and workflows to achieve better performance, lower costs, and higher yields. One important feature of process optimization lies in the optimization of equipment and components that go into tools, which often involves a significant amount of human interaction during both scheduled and unscheduled maintenance activities [22, 35].

1. **Equipment optimization:**

   - **Equipment performance:** Optimization efforts aim to maximize the performance of semiconductor manufacturing equipment, ensuring that they operate at their full capacity and efficiency levels. This involves fine-tuning equipment settings, parameters, and operating conditions to achieve optimal process outcomes [19, 81].

- **Preventive maintenance:** Implementing preventive maintenance schedules and procedures to proactively identify and address potential equipment issues before they escalate into major problems. Regular maintenance helps prevent unexpected downtime and maintains equipment reliability [46, 81].
- **Predictive maintenance:** Leveraging predictive maintenance technologies, such as condition monitoring, predictive analytics, and machine learning algorithms, to anticipate equipment failures and perform maintenance activities proactively. Predictive maintenance minimizes unplanned downtime and reduces maintenance costs.
- **Calibration and alignment:** Ensuring that semiconductor manufacturing equipment is calibrated and aligned correctly to maintain accuracy and consistency in process outcomes. Regular calibration and alignment procedures help optimize equipment performance and minimize process variability.
- **Equipment upgrades and retrofits:** Upgrading equipment with advanced technologies, components, or software updates to enhance performance, increase throughput, and improve process capabilities. Retrofitting older equipment with modern features can extend their lifespan and functionality.

2. **Component optimization:**

- **Tool components:** Optimizing individual components within semiconductor manufacturing tools, such as chambers, valves, pumps, and sensors, to improve reliability, performance, and process control. Upgrading or replacing components with newer, more efficient alternatives can enhance tool functionality and productivity [11].
- **Material selection:** Choosing high-quality materials for critical components to ensure durability, chemical resistance, and compatibility with semiconductor manufacturing processes. Selecting the right materials helps prevent corrosion, contamination, and other issues that can affect process performance.
- **Supply chain management:** Managing the supply chain for equipment components to ensure timely availability of spare parts, consumables, and critical materials. Establishing robust supply chain processes helps minimize lead times, reduce inventory costs, and mitigate the risk of component shortages [12].

3. **Human interaction and maintenance:**

- **Training and skill development:** Providing comprehensive training programs and skill development initiatives for maintenance technicians and operators to ensure they possess the knowledge and expertise required to operate and maintain semiconductor manufacturing equipment effectively.

- **Procedural compliance:** Ensuring that maintenance procedures, protocols, and best practices are followed diligently to maintain equipment integrity, safety, and compliance with industry regulations and standards.
- **Troubleshooting and problem-solving:** Equipping maintenance teams with troubleshooting techniques, diagnostic tools, and problem-solving skills to quickly identify and resolve equipment issues, minimizing downtime and maximizing productivity.
- **Continuous improvement:** Encouraging a culture of continuous improvement among maintenance personnel, where feedback, suggestions, and lessons learned are actively shared and incorporated into maintenance processes to drive ongoing optimization and efficiency gains.

In summary, process optimization in semiconductor manufacturing involves optimizing equipment and components within semiconductor manufacturing tools, as well as human interaction during maintenance activities. By focusing on equipment performance, component optimization, and effective maintenance practices, semiconductor manufacturers can achieve higher levels of productivity, reliability [13], and quality in their manufacturing processes.

## 6.3.1 TPM (Figure 6.1)

**Figure 6.1:** TPM methods.

## TPM – Total Predictive Maintenance ( Minimizing Defects, Minimizing accidents, Minimizing Downtimes)

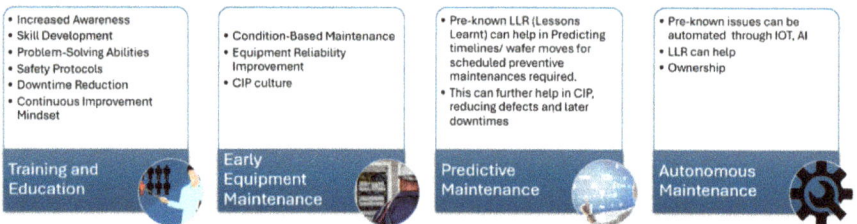

| Training and Education | Early Equipment Maintenance | Predictive Maintenance | Autonomous Maintenance |
|---|---|---|---|
| • Increased Awareness<br>• Skill Development<br>• Problem-Solving Abilities<br>• Safety Protocols<br>• Downtime Reduction<br>• Continuous Improvement Mindset | • Condition-Based Maintenance<br>• Equipment Reliability Improvement<br>• CIP culture | • Pre-known LLR (Lessons Learnt) can help in Predicting timelines/ wafer moves for scheduled preventive maintenances required.<br>• This can further help in CIP, reducing defects and later downtimes | • Pre-known issues can be automated through IOT, AI<br>• LLR can help<br>• Ownership |

### 6.3.2 Training and education

1.  **Increased awareness:** Training programs educate employees about the importance of TPM principles and how their roles contribute to overall equipment effectiveness (OEE). This heightened awareness fosters a culture of ownership and accountability for equipment performance and maintenance.
2.  **Skill development:** Proper training equips employees with the technical skills needed to perform maintenance tasks effectively. This includes equipment inspection, cleaning, lubrication, and minor repairs. Skilled workers can identify potential issues early, preventing defects and breakdowns before they occur.
3.  **Problem-solving abilities:** TPM training often includes modules on problem-solving methodologies such as root cause analysis (RCA) and failure mode and effects analysis (FMEA). By teaching employees how to identify, analyze, and address underlying causes of defects and accidents, training enhances their ability to implement proactive maintenance strategies.
4.  **Safety protocols:** Education on safety protocols and procedures is essential for preventing accidents and injuries in the workplace. Training programs cover topics such as machine guarding, lockout/tagout (LOTO) procedures, hazard identification, and emergency response protocols, ensuring that employees understand and adhere to safety standards.
5.  **Downtime reduction:** Through education on preventive maintenance practices, employees learn how to conduct routine inspections and implement preventive measures to avoid unplanned downtime. Training emphasizes the importance of regular equipment checks, calibration, and predictive maintenance techniques, helping organizations maximize equipment uptime and productivity.
6.  **Continuous improvement mindset:** TPM training instills a culture of continuous improvement, encouraging employees to seek opportunities for optimization and efficiency enhancement. By fostering a mindset of kaizen (continuous improvement), training enables teams to identify areas for improvement and implement sustainable solutions to minimize defects, accidents, and downtime over time [1, 2, 25].

### 6.3.3 Early equipment maintenance

Early equipment maintenance (EEM) is a proactive approach aimed at preventing equipment failures and optimizing performance through timely and systematic maintenance activities. When integrated into total productive maintenance (TPM), EEM plays a vital role in achieving the goals of maximizing equipment effectiveness, minimizing downtime, and reducing defects. Here's how early equipment maintenance contributes to TPM:

1.  **Condition-based maintenance:** In addition to time-based maintenance, EEM incorporates condition-based maintenance strategies that utilize data from equipment sensors, monitoring

systems, and predictive analytics to assess the health and performance of machinery. By monitoring key indicators such as temperature, vibration, and fluid levels, condition-based maintenance enables early detection of abnormalities or deterioration, allowing maintenance interventions to be initiated proactively.

2. **Equipment reliability improvement:** EEM initiatives aim to enhance equipment reliability by addressing common failure modes and implementing measures to mitigate risks. Through techniques such as failure mode and effects analysis (FMEA) and root cause analysis (RCA), maintenance teams identify and prioritize equipment vulnerabilities, designating appropriate maintenance tasks to minimize the likelihood of failures. By proactively addressing reliability issues, EEM contributes to increased equipment uptime and longevity.

3. **Skill development and training:** EEM programs include training and skill development initiatives to equip maintenance technicians with the knowledge and competencies required to perform maintenance tasks effectively. Training covers equipment operation, maintenance procedures, troubleshooting techniques, and safety protocols, empowering employees to identify early warning signs of equipment degradation and take corrective actions promptly.

4. **Integration with autonomous maintenance:** EEM is closely aligned with the autonomous maintenance pillar of TPM, where operators assume responsibility for routine cleaning, inspection, and minor maintenance tasks once indicated through dashboards. By empowering frontline employees to perform basic maintenance activities through "them" taking ownership on dashboards indication. EEM engages the entire workforce in equipment care and fosters a sense of ownership and pride in equipment reliability. Operators play a crucial role in early detection of abnormalities and contribute valuable insights for improving equipment performance [1, 2, 25].

5. **Continuous improvement culture:** EEM encourages a culture of continuous improvement where maintenance processes and practices are regularly reviewed, refined, and optimized. By soliciting feedback from maintenance personnel and leveraging data-driven insights, organizations identify opportunities to streamline maintenance workflows, enhance asset management strategies, and implement best practices for early equipment maintenance. This culture of continuous improvement aligns with the overarching goals of TPM and drives sustainable performance gains over time.

### 6.3.3.1 PM (preventive maintenance) and autonomous maintenance

**Ownership:** EEM is closely aligned with the autonomous maintenance pillar of TPM, where operators assume responsibility for routine cleaning, inspection, and minor maintenance tasks once indicated through dashboards

- Cleaning and inspection: Operators regularly clean equipment surfaces and inspect for signs of wear, damage, or abnormal conditions. This proactive approach helps identify potential issues early and prevents equipment deterioration.

- Tightening and lubricating: Operators check for loose components, fasteners, or belts and tighten them as needed to prevent vibrations and misalignments. They also apply lubricants to moving parts to reduce friction and wear, ensuring smooth operation.
- Minor repairs and adjustments: Operators are trained to perform minor repairs and adjustments to address issues such as misalignments, jams, or minor leaks. By promptly addressing these issues, operators prevent equipment breakdowns and maintain optimal performance.
- Cleaning and restoring standard conditions: Operators follow standardized cleaning procedures to remove dirt, debris, and contaminants from equipment surfaces and surroundings. Restoring equipment to its standard condition helps maintain cleanliness, safety, and operational efficiency.
- Basic troubleshooting and problem-solving: Operators develop troubleshooting skills to identify and resolve common equipment issues, such as sensor malfunctions, clogs, or minor electrical faults. By addressing these issues independently, operators minimize reliance on maintenance technicians and reduce downtime.
- Participation in autonomous maintenance teams: Operators collaborate in autonomous maintenance teams to share knowledge, best practices, and lessons learned. These cross-functional teams work together to improve equipment reliability, efficiency, and safety through continuous improvement initiatives.
- Training and skill development: Operators receive training on equipment operation, maintenance procedures, and safety protocols to build competency and confidence in performing autonomous maintenance activities. Ongoing skill development programs ensure that operators stay updated on the latest maintenance practices and technologies.
- Documentation and data recording: Operators maintain records of autonomous maintenance activities, including inspection results, maintenance tasks performed, and any abnormalities detected. This documentation provides valuable insights for analysis, decision-making, and continuous improvement efforts.

## 6.4 Case Study: Wafer Fabrication Facility – Preventive Maintenance Optimization

**Challenge:** A wafer fabrication facility was experiencing frequent equipment failures and downtime, impacting production schedules and yield rates. Reactive maintenance practices led to increased costs associated with emergency repairs and lost productivity.

**Solution:** The facility implemented a proactive EEM program as part of its total productive maintenance (TPM) initiative. Maintenance schedules were optimized to prioritize preventive tasks such as routine inspections, cleaning, and lubrication. Predictive maintenance techniques, including vibration analysis and thermal imaging, were employed to detect early signs of equipment degradation.

**Outcome:** By adopting preventive maintenance optimization, the facility achieved significant improvements in equipment reliability and uptime. Downtime due to unexpected failures was minimized, leading to smoother production operations and higher yield rates. The proactive approach to maintenance resulted in cost savings by reducing the need for emergency repairs and extending the lifespan of critical equipment.

## 6.5 Case Study: Semiconductor Manufacturing Plant – Condition-based Monitoring Implementation

**Challenge:** A semiconductor manufacturing plant faced challenges in maintaining equipment reliability and meeting stringent quality standards. Variability in process parameters and occasional equipment failures led to yield losses and product defects.

**Solution:** To enhance equipment monitoring and diagnostics, the plant implemented a condition-based monitoring (CBM) system. Sensors were installed on key equipment to continuously monitor variables such as temperature, pressure, and flow rates. Advanced analytics software was used to analyze sensor data and predict potential failure modes.

**Outcome:** With the implementation of CBM, the plant gained real-time insights into equipment health and performance. Early warnings of impending failures allowed maintenance teams to proactively address issues before they caused production disruptions. As a result, the plant experienced improvements in yield rates, product quality, and overall operational efficiency.

## 6.6 Case Study: Integrated Circuit Packaging Facility – Operator-led Maintenance Initiative

**Challenge:** An integrated circuit packaging facility was grappling with equipment downtime and maintenance backlog, affecting throughput and delivery schedules. The maintenance team was stretched thin, struggling to keep up with reactive repair tasks.

**Solution:** To alleviate the burden on the maintenance team and foster a culture of equipment ownership, the facility launched an operator-led maintenance initiative. Operators received training on equipment care and were empowered to perform routine maintenance tasks such as cleaning, minor adjustments, and lubrication.

**Outcome:** The operator-led maintenance initiative led to a reduction in equipment downtime and improved overall equipment effectiveness (OEE). Operators became more engaged in equipment upkeep and were able to identify and address minor issues before they escalated. The collaborative approach between operators and maintenance technicians resulted in a more efficient and proactive maintenance workflow, contributing to increased productivity and equipment reliability.

# 7

## Overview of AI and Applications in Semiconductor Manufacturing

Artificial intelligence (AI) can be incorporated into various aspects of process optimization in semiconductor manufacturing to enhance efficiency, productivity, and quality. Here are some key areas where AI can be applied:

1. **Predictive maintenance [20, 21, 22]**

   - AI algorithms can analyze historical equipment data, sensor readings, and operational parameters to predict equipment failures and maintenance needs accurately.
   - By identifying patterns and trends in equipment performance, AI can forecast potential issues and schedule maintenance activities proactively, minimizing unplanned downtime and reducing maintenance costs.
   - Predictive maintenance models can optimize maintenance schedules based on factors such as equipment usage, environmental conditions, and production demands, maximizing equipment uptime and availability.

2. **Process control and optimization [21]**

   - AI-powered process control systems can monitor and adjust manufacturing processes in real-time to optimize parameters such as temperature, pressure, and chemical concentrations.
   - Machine learning algorithms can analyze complex process data to identify opportunities for optimization, improve yield rates, and reduce cycle times.

- AI-based optimization models can dynamically adjust process settings and parameters to achieve desired outcomes, such as improving product quality, reducing defects, and minimizing variation.

3. **Anomaly detection and fault diagnosis [22]**

- AI algorithms can detect anomalies and deviations in equipment performance, process parameters, and product quality by comparing real-time data with expected patterns and standards.
- Machine learning techniques, such as anomaly detection algorithms and fault classification models, can identify root causes of issues and recommend corrective actions to maintenance teams.
- AI-based fault diagnosis systems can diagnose equipment malfunctions and process abnormalities more accurately and rapidly than traditional manual methods, reducing diagnosis time and improving troubleshooting efficiency.

4. **Supply chain optimization [23]**

- AI-driven supply chain management systems can analyze historical data, demand forecasts, and inventory levels to optimize procurement, logistics, and inventory management processes.
- Machine learning algorithms can predict component shortages, lead times, and supplier risks, allowing manufacturers to proactively address supply chain disruptions and mitigate risks.
- AI-based optimization models can optimize supply chain workflows, minimize inventory costs, and improve supplier relationships by identifying opportunities for cost reduction and process improvement.

5. **Human–machine collaboration [24]**

- AI technologies can augment human capabilities in maintenance, troubleshooting, and decision-making by providing real-time insights, recommendations, and assistance.
- AI-powered diagnostic tools and augmented reality interfaces can assist maintenance technicians in diagnosing equipment issues, accessing repair procedures, and performing tasks more efficiently.
- Collaborative robots (cobots) equipped with AI capabilities can work alongside human operators to automate repetitive tasks, enhance safety, and improve overall productivity in semiconductor manufacturing environments [20, 21, 22, 26].

## 7.1 How to Incorporate AI/IOT into Various Aspects of Process Optimization

### 7.1.1 Data collection and integration [25]

- Implement sensors and data collection systems throughout the manufacturing process to capture equipment data, process parameters, and product quality metrics.
- Integrate data from disparate sources, such as manufacturing equipment, MES (manufacturing execution system), SCADA (supervisory control and data acquisition) systems, and enterprise databases, into a centralized data repository.

### 7.1.2 Data preprocessing and cleaning

- Preprocess and clean raw data to remove noise, outliers, and inconsistencies, ensuring the accuracy and reliability of the data used for AI modeling and analysis.
- Apply data transformation techniques, such as normalization, scaling, and feature engineering, to prepare data for AI algorithms [26].

### 7.1.3 AI model development

- Develop AI models [27], such as machine learning algorithms, deep learning neural networks, and statistical models, tailored to specific applications in semiconductor manufacturing.
- Select appropriate AI techniques based on the nature of the problem, available data, and desired outcomes, considering factors such as predictive maintenance, process optimization, anomaly detection, and supply chain management.

### 7.1.4 Training and validation

- Train [28] AI models using historical data and domain expertise to learn patterns, correlations, and relationships between input variables and target outcomes.
- Validate and fine-tune AI models using techniques like cross-validation, hyperparameter tuning, and ensemble methods to ensure robustness and generalization performance.

### 7.1.5 Real-time monitoring and control

- Deploy AI models in real-time monitoring and control [29] systems to continuously analyze streaming data and make proactive decisions to optimize equipment performance, process parameters, and product quality.

- Implement closed-loop control systems that automatically adjust process settings and parameters based on AI insights and recommendations, minimizing manual intervention and maximizing efficiency [24].

### 7.1.6 Continuous improvement and iterative learning

- Establish mechanisms for continuous improvement and iterative learning [30], where AI models are updated and refined over time based on feedback, new data, and evolving manufacturing requirements.
- Foster a culture of innovation and experimentation, encouraging collaboration between data scientists, engineers, and domain experts to explore new AI techniques and applications in semiconductor manufacturing.

## 7.2 Integration of Edge Computing and Fog computing with IoT [31]

Edge computing and fog computing are both paradigms that extend the capabilities of cloud computing to the edge of the network, enabling data processing and analysis closer to the data source. While they share similarities, they differ in their scope, architecture, and deployment models.

### 7.2.1 Edge computing

Edge computing [31] refers to the practice of processing data near the edge of the network, closer to where it is generated, rather than sending it to a centralized data center or cloud environment. In edge computing, data processing tasks are performed on local devices, such as routers, gateways, or edge servers, situated close to the data source. This approach reduces latency, bandwidth usage, and dependence on centralized infrastructure, making it ideal for applications that require real-time processing, low latency, and efficient use of network resources. Edge computing is commonly used in IoT (Internet of Things) deployments, industrial automation, autonomous vehicles, and other latency-sensitive applications.

### 7.2.2 Fog computing

Fog computing [31] extends the principles of edge computing by introducing a hierarchical architecture that incorporates intermediary computing nodes, known as fog nodes or fog devices, between the edge devices and the centralized cloud infrastructure. Fog nodes are typically deployed at the network edge, such as within access points, base stations, or switches, and provide computing, storage, and networking capabilities to support edge applications. Fog computing enables distributed data processing and analysis across a network of fog nodes, allowing for scalable and efficient execution of edge computing tasks. This architecture is well-suited for applications that require distributed computing, data aggregation, and coordination among edge devices, such as smartsmart cities, intelligent transportation systems, and healthcare monitoring solutions.

- Utilize edge computing technologies to deploy lightweight AI models directly on manufacturing equipment or edge devices, enabling real-time analysis and decision-making at the point of data generation.
- Integrate AI-powered IoT (Internet of Things) devices and sensors into semiconductor manufacturing equipment to collect and transmit data for AI analysis and control, enabling seamless connectivity and interoperability [31].

## 7.2.3 Differences between edge computing, fog computing

**Architecture:** Edge computing typically involves direct processing of data on local devices at the network edge, without intermediate nodes.

Fog computing introduces intermediary fog nodes between the edge devices and the centralized cloud infrastructure, enabling distributed processing and coordination.

**Scope:** Edge computing focuses on processing data locally at the edge of the network, primarily to reduce latency and improve responsiveness for edge applications.

Fog computing extends the capabilities of edge computing by providing a distributed computing infrastructure that spans across multiple edge devices and fog nodes, enabling scalable and collaborative edge computing tasks.

**Deployment models:** Edge computing deployments involve deploying computing resources directly on edge devices or within close proximity to them, such as on-premises servers, gateways, or routers.

Fog computing deployments entail deploying fog nodes at strategic locations within the network, typically at the network edge, to provide intermediary computing and storage capabilities for edge applications [27, 28, 29].

### 7.2.4 Opportunities for process optimization using edge computing and fog computing

Edge computing and fog computing [31] offer several opportunities for process optimization in semiconductor manufacturing. Here's how they can be utilized:

1. **Real-time data processing**

   - Edge computing enables real-time data processing and analysis directly on manufacturing equipment or edge devices, allowing for immediate insights and decision-making without relying on centralized cloud infrastructure.
   - Fog computing extends this capability by distributing computing resources across fog nodes located at strategic points in the manufacturing environment, enabling distributed data processing and analysis closer to the data source.

2. **Predictive maintenance**

   - Edge computing can be used to deploy predictive maintenance models directly on manufacturing equipment, enabling real-time monitoring of equipment health and performance.
   - Fog computing extends this capability by aggregating data from multiple edge devices and applying machine learning algorithms to predict equipment failures and maintenance needs across the manufacturing facility.

3. **Process control and optimization**

   - Edge computing allows for localized process control and optimization by implementing closed-loop control systems directly on manufacturing equipment, adjusting process parameters in real-time based on sensor data and performance metrics.
   - Fog computing extends this capability by coordinating process control and optimization tasks across multiple edge devices and fog nodes, enabling collaborative decision-making and resource allocation.

4. **Quality assurance**

   - Edge computing enables real-time quality assurance by performing in-line inspection and analysis of semiconductor manufacturing processes directly on the production line, identifying defects and anomalies early in the manufacturing process.

- Fog computing extends this capability by integrating data from multiple edge devices and fog nodes to perform comprehensive quality assurance across the entire manufacturing environment, ensuring consistency and reliability in product quality.

5. **Supply chain optimization**

- Edge computing can be used to optimize supply chain processes by deploying AI-based inventory management systems directly at manufacturing facilities, predicting component shortages, and optimizing inventory levels in real-time.
- Fog computing extends this capability by coordinating supply chain optimization tasks across multiple manufacturing sites and supply chain partners, enabling end-to-end visibility and coordination in the semiconductor manufacturing supply chain.

6. **Human–machine collaboration**

- Edge computing facilitates human–machine collaboration by deploying AI-driven decision support systems and augmented reality interfaces directly on manufacturing equipment, assisting operators and maintenance technicians in troubleshooting, repair, and decision-making tasks.
- Fog computing extends this capability by providing a distributed computing infrastructure that supports collaborative human–machine interactions across multiple edge devices and fog nodes, enabling seamless integration of AI-driven tools and technologies into the manufacturing workflow [24, 25, 26].

# 8

# Overview of IoT Technologies and Applications

The Internet of Things (IoT) has emerged as a revolutionary paradigm, connecting everyday objects and devices to the internet, enabling them to collect, exchange, and analyze data autonomously. This overview explores the fundamental concepts, technologies, and diverse applications of IoT across various domains [24].

1. **Fundamental concepts of IoT**

   - IoT encompasses a network of interconnected devices, sensors, actuators, and systems that communicate and collaborate seamlessly over the internet.
   - These IoT devices collect data from their surroundings through sensors and transmit it to centralized platforms or other connected devices for processing and analysis.
   - Communication protocols such as Wi-Fi, Bluetooth, Zigbee and cellular networks enable seamless connectivity and data exchange between IoT devices and the cloud [32]

2. **Key technologies driving IoT**

   - Sensor technology: IoT devices are equipped with a variety of sensors, including temperature sensors, humidity sensors, motion sensors, GPS receivers, and accelerometers, to collect real-time data from the physical world [34].
   - Connectivity solutions: IoT devices utilize diverse connectivity options such as Wi-Fi, Bluetooth, RFID, NFC, Zigbee, cellular networks (2G, 3G, 4G, and 5G), and satellite communication to establish network connections and transmit data [33].

- Edge computing: Edge computing enables data processing and analysis to be performed closer to the data source, reducing latency and bandwidth usage. Edge devices preprocess data locally before transmitting it to the cloud for further analysis.
- Cloud computing: Cloud platforms provide scalable storage, computing power, and analytics capabilities for processing and analyzing vast amounts of IoT data. Cloud-based IoT platforms enable centralized management, monitoring, and control of distributed IoT deployments.
- Machine learning and artificial intelligence: AI and ML algorithms analyze IoT data to extract actionable insights, predict future trends, and automate decision-making processes. These technologies enable predictive maintenance, anomaly detection, and optimization of IoT systems and applications.

3. **Applications of IoT across industries**

- Smart cities: IoT technologies are used to monitor and manage urban infrastructure, including smartsmart street lighting, waste management, traffic control, parking management, environmental monitoring, and public safety [27, 28, 29].
- Industrial IoT (IIoT): In industrial settings, IoT enables remote monitoring, predictive maintenance, asset tracking, and optimization of manufacturing processes. IIoT applications include condition-based monitoring, supply chain management, energy management, and real-time production control.
- Healthcare: IoT devices and wearables monitor patients' health metrics, track medication adherence, facilitate remote patient monitoring, and enable telemedicine services. IoT-based healthcare solutions improve patient outcomes, enhance efficiency, and reduce healthcare costs.
- Agriculture: IoT sensors monitor soil moisture levels, environmental conditions, crop health, and livestock behavior to optimize agricultural operations, increase crop yields, and conserve resources. Smart farming solutions leverage IoT for precision agriculture, automated irrigation, and livestock management [33].
- Smart homes: IoT [32] devices automate and control various aspects of home environments, including smartsmart thermostats, lighting systems, security cameras, smart appliances, and voice-controlled assistants. Smart home solutions enhance convenience, comfort, and energy efficiency for homeowners.

4. **Challenges and considerations [27]**

- Security and privacy: IoT devices are vulnerable to cyberattacks, data breaches, and privacy violations. Implementing robust security measures, encryption protocols, and access controls is essential to safeguard IoT deployments.

- Interoperability: Ensuring compatibility and interoperability bet- ween diverse IoT devices, protocols, and platforms is crucial for seamless integration and data exchange in IoT ecosystems.
- Scalability and reliability: Scaling IoT deployments to accommodate large numbers of devices and data volumes while maintaining reliability, performance, and data integrity poses challenges in terms of network bandwidth, processing power, and infrastructure scalability.
- Data management and analytics: Managing and analyzing vast amounts of IoT data poses challenges related to data storage, processing, latency, and real-time analytics. Adopting efficient data management strategies and scalable analytics solutions is essential to derive actionable insights from IoT data.

## 8.1 Integration of IoT Technologies in Semiconductor Manufacturing

### 8.1.1 Smart manufacturing and predictive maintenance

- IoT [32] sensors can be deployed throughout semiconductor fabrication facilities to monitor equipment health, process parameters, and environmental conditions in real-time.
- Predictive maintenance models powered by IoT data analytics can anticipate equipment failures, detect abnormalities, and schedule maintenance proactively to prevent costly downtime.
- Condition-based monitoring of semiconductor manufacturing equipment enables early detec- tion of anomalies, allowing for timely intervention and optimization of equipment performance.

### 8.1.2 Quality control and process optimization

- IoT-enabled sensors collect data on wafer quality, process variability, and defect rates at various stages of semiconductor manufacturing.
- Real-time data analytics and machine learning algorithms analyze IoT data to identify trends, patterns, and correlations, enabling continuous quality improvement and process optimization.
- Adaptive control systems adjust manufacturing parameters dynamically based on IoT insights, ensuring consistent product quality and minimizing defects in semiconductor production.

### 8.1.3 Supply chain management and inventory control

- IoT devices track inventory levels, materials usage, and logistics in semiconductor manufactur- ing supply chains.
- RFID tags and IoT sensors enable real-time tracking and tracing of semiconductor components, raw materials, and finished products throughout the production process.

- Demand forecasting models powered by IoT data analytics optimize inventory management, procurement, and production planning, reducing lead times and minimizing supply chain disruptions.

### 8.1.4 Energy efficiency and sustainability

- IoT sensors monitor energy consumption, environmental impact, and resource usage in semiconductor fabrication facilities.
- Energy management systems leverage IoT data to identify energy-intensive processes, optimize equipment utilization, and implement energy-saving measures.
- Sustainability initiatives driven by IoT technologies enable semiconductor manufacturers to reduce carbon emissions, minimize waste generation, and achieve environmental compliance goals.

### 8.1.5 Real-time process control and automation

- IoT-connected devices and actuators enable remote monitoring, control, and automation of semiconductor manufacturing processes.
- Edge computing platforms preprocess IoT data locally and execute control algorithms in real-time, minimizing latency and improving responsiveness.
- Closed-loop control systems leverage IoT insights to adjust process parameters dynamically, ensuring optimal performance, yield, and efficiency in semiconductor production.

### 8.1.6 Worker safety and ergonomics

- IoT wearables and safety devices enhance worker safety and ergonomics in semiconductor manufacturing environments.
- Smart sensors monitor environmental conditions, air quality, and ergonomic factors to prevent accidents, minimize exposure to hazardous substances, and improve worker well-being.
- IoT-enabled safety systems trigger alerts and emergency responses in case of incidents or abnormal conditions, ensuring rapid intervention and mitigating risks to personnel.

By integrating IoT technologies into semiconductor manufacturing processes, organizations can unlock new opportunities for efficiency, quality improvement, and innovation. From predictive maintenance and real-time process control to supply chain optimization and sustainability initiatives, IoT-driven solutions offer transformative benefits for semiconductor manufacturers seeking to enhance competitiveness and achieve operational excellence [27, 28, 29].

# 9

---

# Fundamentals of Semiconductor Defects

---

When we talk about a 3000 mm wafer in semiconductor fabrication, we are referring to the size of the silicon wafer used as the substrate for manufacturing integrated circuits (ICs) and other semiconductor devices. The notation "3000 mm" typically represents the diameter of the wafer.

In the semiconductor industry, the most common wafer sizes are 200 mm (8 inches), 300 mm (12 inches), and, more recently, 450 mm (18 inches). The larger wafer sizes, such as 300 mm, are used to increase the number of chips that can be produced on a single wafer, leading to improved manufacturing efficiency and reduced production costs per chip. The transition from smaller wafer sizes (e.g., 200 mm) to larger ones (e.g., 300 mm) has been a significant trend in semiconductor manufacturing. Larger wafers allow for more chips to be fabricated simultaneously, which can result in economies of scale and improved yields. However, the transition to even larger wafer sizes, such as 450 mm, has been more challenging due to technological and economic factors. Thus, with increased chip quantity, there needs to be effective qualitative measurements to ensure chips are manufactured defect free.

It's worth noting that the size of the wafer can impact the overall production process, equipment, and cost structure within a semiconductor fabrication facility. The choice of wafer size is influenced by considerations such as cost-effectiveness, manufacturing efficiency, and the technological capabilities of the semiconductor industry at a given time.

Semiconductor defects can arise during various stages of the semiconductor manufacturing process, and they can significantly impact the performance, reliability, and yield of semiconductor devices. Here are some fundamental aspects of semiconductor defects:

## 9.1 Types of Defects

- **Point defects:** These defects [35, 36, 37, 38, 39, 40, 45] occur at specific locations within the semiconductor lattice and include vacancies (missing atoms), interstitials (extra atoms inserted into the lattice), and impurities (foreign atoms incorporated into the lattice).
- **Line defects:** These defects involve dislocations or line imperfections in the crystal lattice structure, which can act as sites for localized strain or stress accumulation. Example: Dislocation – a line defect where an extra row of atoms is inserted into the lattice.
- **Surface defects:** These defects occur at the semiconductor surface and include roughness, contamination, and oxidation, which can affect surface properties and device performance. Example: Scratch – physical damage on the surface of the silicon wafer.
- **Bulk defects:** These defects occur within the bulk volume of the semiconductor material and can include crystallographic defects such as stacking faults, twins, and grain boundaries.
- **Process-induced defects:** These are defects introduced during the various processing steps, such as photolithography, etching, and deposition. Example: Leftover photoresist material after the etching process.

Overall, defects in semiconductor manufacturing can be easily classified into two categories: intrinsic and extrinsic. Extrinsic defects are more common than intrinsic defects in semiconductor manufacturing processes. In the event of a chip failing in functionality, the probability that it might be due to an extrinsic defect is higher compared to that of intrinsic defects.

1. **Intrinsic defects**

   - **Definition:** Intrinsic defects are defects that originate from within the material itself, typically due to deviations in the arrangement of atoms or molecules.
   - **Cause:** These defects are caused by factors inherent to the material, such as impurities, vacancies, or lattice imperfections.
   - **Examples:** Examples of intrinsic defects include vacancies (missing atoms in the crystal lattice), interstitials (extra atoms squeezed into the lattice), and dislocations (defects in the crystal structure).
   - **Impact:** Intrinsic defects can affect the material's properties, such as its electrical conductivity, mechanical strength, and optical properties.

2. **Extrinsic defects**

   - **Definition:** Extrinsic defects are defects introduced into the material from external sources or environmental factors.
   - **Cause:** These defects are caused by factors external to the material, such as contamination during manufacturing processes, exposure to radiation, or chemical reactions with impurities.

- **Examples:** Examples of extrinsic defects include impurities introduced during fabrication, surface contamination, or defects induced by external stress or mechanical damage.
- **Impact:** Extrinsic defects can also alter the material's properties and performance, often leading to changes in conductivity, optical transparency, or mechanical behavior.

## 9.2 Sources of Defects

- **Crystal growth:** Defects can originate during the crystal growth process, such as dislocations introduced during solidification or impurities incorporated from the growth environment.
- **Wafer fabrication:** Defects can be introduced during wafer fabrication processes, including thin film deposition, photolithography, etching, doping, and annealing.
- **Packaging and assembly:** Defects can arise during packaging and assembly processes, such as die attachment, wire bonding, encapsulation, and testing [26, 35, 36, 37, 38, 39, 40, 45].

## 9.3 Effects of Defects

- **Electrical effects:** Defects [36] can alter the electrical properties of semiconductor devices, leading to increased leakage currents, decreased carrier mobility, and reduced device performance.
- **Reliability issues:** Defects can compromise the reliability of semiconductor devices, leading to premature device failure, degraded performance over time, and susceptibility to environmental stresses such as temperature, humidity, and radiation.
- **Yield loss:** Defects can result in lower production yields, as defective devices may fail to meet quality standards or performance specifications, leading to increased manufacturing costs and reduced profitability.
- **Functional failures:** Defects can cause functional failures in semiconductor devices, leading to malfunctions, errors, or non-operational behavior in electronic systems and applications.

## 9.4 Detection and Characterization

- **Metrology techniques:** Various metrology techniques are used to detect and characterize [37] semiconductor defects, including scanning electron microscopy (SEM), transmission electron microscopy (TEM), atomic force microscopy (AFM), X-ray diffraction (XRD), and optical microscopy.
- **Electrical testing:** Electrical testing methods, such as current–voltage ($IV$) measurements, capacitance–voltage ($CV$) measurements, and defect analysis techniques, can be used to identify and analyze electrical defects in semiconductor devices.
- **Non-destructive testing:** Non-destructive testing methods, such as acoustic microscopy, infrared thermography, and laser-based techniques, can provide insights into internal defects and material properties without damaging the semiconductor device.

## 9.5 Defect Engineering and Mitigation

- **Process optimization [38]:** Semiconductor manufacturers employ process optimization techniques to minimize defect formation during fabrication processes, including optimizing process parameters, improving material purity, and enhancing process control.
- **Defect engineering:** Defect engineering techniques, such as intentional doping, annealing, and stress engineering, can be used to manipulate defect properties and mitigate their effects on device performance.
- **Quality assurance:** Implementing robust quality assurance measures, such as in-line inspection, process monitoring, and statistical process control (SPC), can help identify and address defects early in the manufacturing process, ensuring high product quality and yield.

## 9.6 Most Common Defects in Semiconductor Manufacturing

Common defects in semiconductor manufacturing, often associated with equipment failure or processing errors, include those listed below. To substantiate this assertion, hypothetical case studies are provided, illustrating that whenever a failure analysis is conducted on a chip, the objective is to identify corrective and preventive actions through root cause analysis (RCA). The 8D methodology is typically employed, which entails eight disciplines for problem-solving. Additionally, methodologies such as DMAIC, PDCA, and the "five whys" approach are commonly utilized. During the implementation of corrective actions and preventive measures, the aim is to institute long-term solutions. This often necessitates revisiting the foundational aspects of semiconductor manufacturing, including the processes involved, materials, and equipment utilized [39].

1. **Particle contamination:** Occurs when foreign particles, such as dust or residues, are introduced during the manufacturing process. This can lead to defects in the semiconductor devices, affecting their performance and reliability. Usually, when a wafer after processing has issues chucking on the wafer platform, also known as the wafer chuck, there is a possibility of wafer contamination through particle presence (Figure 9.1). This assumption can be verified by visually looking at the wafer or putting the wafer for IR scanning. In case particle presence is detected, it is the duty of the engineer to check if the wafer can be cleaned without damaging the critical dimensions and wafer functionality in terms of electrical properties. Then processing should be resumed per normal, else the wafer has to be scrapped for not meeting the requirements of processing due to particle contamination [37, 38, 39, 40].
   Implementing a shift left mechanism involves early detection and mitigation of particle contamination through rigorous monitoring of cleanroom conditions and equipment maintenance [40]

**Figure 9.1:** An example of possible particle contamination on wafer.

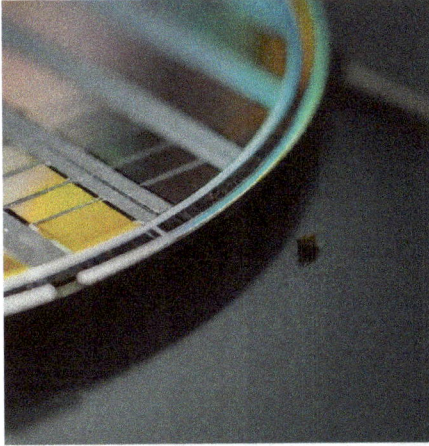

2. **Micro-arcing:** This defect arises from electrical discharge between conductive surfaces in semiconductor equipment (Figure 9.2). It can result in localized damage to the semiconductor material, affecting device functionality.

   The cause of micro-arcing can be electrostatic discharge. There can be multiple causes, such as metal chipping inside the chamber while processing, causing plasma to attack on metal surface generating ESD, thereby causing micro-arcing on the wafer.

   Implementing shift left practices involves improving equipment design and maintenance procedures to minimize the risk of micro-arcing and ensure proper electrical isolation [41].

**Figure 9.2:** An Example of the effect of micro-arcing on a wafer.

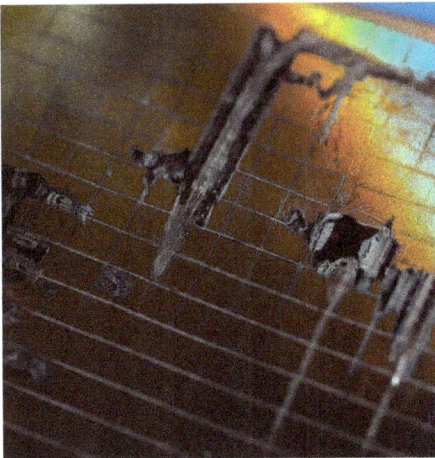

3. **Etching inconsistencies:** Etching processes are used to selectively remove material from semiconductor wafers to create desired patterns. Inconsistencies in etching (Figure 9.3) can lead to uneven surfaces or improper feature dimensions, causing defects in the final devices. Implementing shift left strategies involves optimizing etching parameters, regularly calibrating equipment, and employing real-time process monitoring to detect and address etching inconsistencies early in the manufacturing process [42].

**Figure 9.3:** An example of the effect of etching inconsistencies on a wafer.

4. **Arcing:** Both arcing and micro-arcing involve the phenomenon of electrical discharge, but they differ in scale, intensity, and consequences (Figures 9.4 and 9.5):

   - **Larger scale:** Arcs are typically visible to the naked eye, ranging from millimeters to centimeters in length.
   - **Higher intensity:** Arcs involve a more significant flow of current, often exceeding several amperes.
   - **Greater damage:** Arcs generate higher heat and can cause significant damage to surrounding materials, including melting, burning, and vaporization.
   - **Common causes:** Arcs can occur due to various factors like high voltage breakdown, loose connections, or insulation failure.

**Figure 9.4:** An example of major arcing on the IC.

**Figure 9.5:** A example of the effect of arcing on a wafer.

5. **Micro-arcing:**

- **Smaller scale:** Micro-arcs are microscopic in size, often undetectable by the naked eye, typically ranging from micrometers to tens of micrometers.
- **Lower intensity:** Micro-arcs involve a much smaller flow of current, typically in the milliampere or microampere range.
- **Minor damage:** While micro-arcs can cause localized damage, it's usually less severe compared to regular arcing, potentially affecting surface properties or inducing localized heating.
- **Common causes:** Micro-arcing often occurs due to factors like surface contamination, microscopic protrusions, or weak electrical insulation in close proximity.

Table 9.1 summarizes the key differences.

**Table 9.1:** Differences between arcing and micro-arcing.

| Feature | Arcing | Micro-arcing |
|---|---|---|
| Scale | Visible to the naked eye | Microscopic |
| Current flow | High (amps) | Low (milliamps or microamps) |
| Damage | Significant | Minor (localized) |
| Common causes | High voltage breakdown, loose connections, insulation failure | Contamination, microscopic protrusions, weak insulation |

In the context of semiconductor manufacturing, both arcing and micro-arcing can pose challenges:

- **Arcing:** Can damage equipment, cause fires, and disrupt production processes.
- **Micro-arcing:** Can degrade device performance, introduce defects, and reduce device reliability.

Therefore, it's crucial to implement preventive measures to minimize the risk of both types of arcing in semiconductor manufacturing environments. This may involve:

- Maintaining proper equipment cleanliness and insulation.
- Monitoring electrical connections and voltage levels.
- Employing proper grounding and shielding techniques.
- Using specialized equipment and techniques to detect and address micro-arcing events.

6.  **Lithography errors**: Errors in the lithography [43] process, such as misalignment of masks or exposure dose variations, can result in pattern distortion or feature inaccuracies on the semiconductor wafers (Figure 9.6). Implementing shift left approaches involves using advanced lithography techniques, conducting thorough equipment maintenance, and implementing automated inspection systems to detect and correct lithography errors before they impact device yield.

**Figure 9.6:** An example of the effect of a lithography error on a wafer.

7.  **Deposition non-uniformity:** Deposition processes are used to deposit thin films of materials onto semiconductor wafers. Non-uniform deposition can lead to variations in film thickness across the wafer, compromising device performance and reliability (Figure 9.7). Implementing shift left methodologies involves optimizing deposition parameters, ensuring proper equipment calibration, and implementing in-line monitoring systems to detect and address deposition non-uniformities early in the manufacturing process [44].

**Figure 9.7:** An example of the effect of deposition non-uniformity on a wafer.

8.  **Chemical or metal contamination:** Chemical contamination can occur when impurities are introduced into processing chemicals or when cleaning solutions are not properly filtered or replenished (Figures 9.8 and 9.9). Contamination can lead to defects such as metal ion contamination or chemical residue buildup on semiconductor surfaces, affecting device performance and reliability. Shift left approaches involve strict adherence to chemical handling protocols, regular monitoring of chemical purity, and implementing preventive maintenance schedules for chemical delivery systems [45].

**Figure 9.8:** An example of chemical contamination on a wafer.

**Figure 9.9:** An Example of metal contamination on a wafer.

## 9.7 Process-Related Defects and Equipment Troubleshooting: A Hypothetical Case Study

**Background:** A semiconductor fabrication facility (fab) specializing in the production of advanced microprocessors experienced an increase in defect rates during the fabrication of a new product line. Defects included pattern irregularities, oxide thickness variations, and dopant concentration inconsistencies, leading to yield losses and increased production costs [1, 2].

**Investigation:** The fab's engineering team conducted a thorough investigation to identify the root causes of the defects. Process monitoring data and equipment logs revealed anomalies coinciding with the occurrence of defects. Further analysis indicated that the defects were predominantly occurring in wafers processed by specific equipment clusters, particularly during critical lithography and deposition steps.

**Equipment troubleshooting:** The engineering team focused on troubleshooting the identified equipment clusters to pinpoint the sources of the defects. They conducted comprehensive equipment diagnostics, including calibration checks, alignment verification, and inspection of critical components such as optics, nozzles, and deposition chambers. Several issues were identified, including misalignment of lithography masks, deposition nozzle clogging, and temperature fluctuations in annealing chambers[11].

**Resolution:** Based on the findings from equipment troubleshooting, corrective actions were implemented to address the identified issues. This included realigning lithography masks, cleaning deposition nozzles, and implementing tighter temperature control measures during annealing. Additionally, preventive maintenance schedules were revised to include more frequent checks and calibration procedures for critical equipment components.

**Outcome:** Following the implementation of corrective actions, the fab observed a significant reduction in defect rates and improvement in overall yield for the new product line. Process stability and equipment reliability were enhanced, leading to smoother production runs and reduced downtime. The success of the equipment troubleshooting efforts underscored the importance of proactive equipment maintenance and process optimization in mitigating process-related defects in semiconductor manufacturing.

**Key takeaways:** As we shift left and delve deeper into the root causes by asking multiple "whys" (using techniques such as the five whys or root cause analysis), we uncover that process-related defects in semiconductor manufacturing frequently stem from equipment issues. This underscores the pivotal importance of equipment reliability and maintenance in the semiconductor industry [71].

Thorough equipment troubleshooting and diagnostics are essential for identifying root causes of defects and implementing effective corrective actions.

Proactive maintenance and optimization of equipment parameters are vital for ensuring process stability, yield improvement, and overall fab performance.

This case study illustrates how semiconductor manufacturers can address process-related defects by focusing on equipment troubleshooting and proactive maintenance practices, ultimately improving product quality and production efficiency.

## 9.8 Process-related Defects in Dry Etch Semiconductor Manufacturing: A Hypothetical Case Study

**Background:** A semiconductor fabrication facility specializing in the production of advanced memory devices experienced an increase in defect rates during the dry etch process step. Defects included etch depth variations, pattern fidelity

**Figure 9.10:** Case study process-related defects.

issues, and sidewall roughness, leading to yield losses and reduced device performance (Figure 9.10).

**Investigation:** The fab's engineering team initiated an investigation to identify the root causes of the defects occurring during the dry etch process. Process monitoring data and defect maps revealed localized defects clustered in specific regions of the wafers processed by certain dry etch tools. Equipment logs indicated variations in process parameters such as gas flow rates, chamber pressure, and RF power during defect occurrences.

**Equipment troubleshooting:** The engineering team focused on troubleshooting the identified dry etch tools to pinpoint the sources of the defects. They conducted comprehensive equipment diagnostics, including chamber cleaning, gas flow calibration, RF power tuning, and inspection of critical components such as electrodes and gas distribution plates. An in-depth analysis revealed issues such as electrode erosion, gas line contamination, and chamber particle buildup affecting process uniformity and etch profile control.

**Resolution:** Based on the findings from equipment troubleshooting, corrective actions were implemented to address the identified issues. This included replacing worn electrodes, purging contaminated gas lines, and implementing enhanced chamber cleaning protocols. Process parameters were optimized to ensure consistent etch rates and sidewall profiles across the wafer surface. Additionally, preventive maintenance schedules were revised to include more frequent chamber inspections and component replacements.

**Outcome:** Following the implementation of corrective actions, the fab observed a significant reduction in defect rates and improvement in etch process uniformity and repeatability. EtchEtch depth variations and sidewall roughness were minimized, leading to enhanced device performance and yield. The success of the equipment troubleshooting efforts demonstrated the critical role of proactive equipment maintenance and process optimization in mitigating process-related defects in dry etch semiconductor manufacturing.

**Key takeaways:**

1. Process-related defects in dry etch semiconductor manufacturing can be attributed to equipment issues such as electrode erosion, gas line contamination, and chamber particle buildup.
2. Thorough equipment troubleshooting and diagnostics are essential for identifying root causes of defects and implementing effective corrective actions.
3. Proactive maintenance and optimization of equipment parameters are vital for ensuring process stability, yield improvement, and overall fab performance in dry etch processes.

This case study highlights the importance of equipment troubleshooting and proactive maintenance practices in addressing process-related defects in dry etch semiconductor manufacturing, ultimately leading to improved product quality and production efficiency.

## 9.9 Equipment-related Process Defects in Chemical Vapor Deposition (CVD): A Hypothetical Study

**Background:** A semiconductor fabrication facility utilizing CVD processes for thin film deposition experiences an increase in defect rates, including film non-uniformity and contamination issues (Figure 9.11).

**Investigation:** The engineering team conducts equipment diagnostics and identifies irregularities in gas flow patterns and temperature control within the CVD chambers.

**Root cause analysis:** The defects are traced back to malfunctioning gas distribution systems and inadequate chamber cleaning procedures.

**Figure 9.11:** Equipment-related process defects in chemical vapor deposition (CVD).

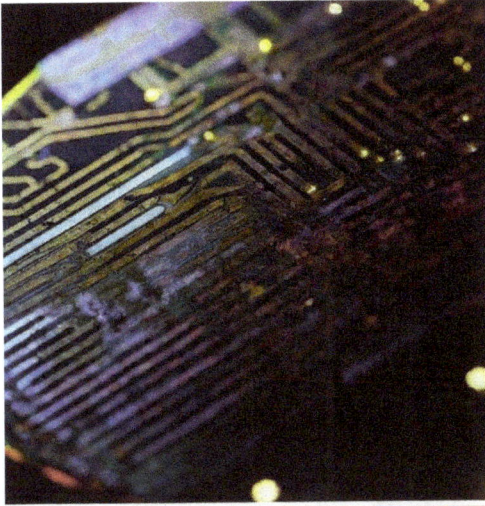

**Corrective action:** The team implements maintenance protocols to recalibrate gas flow controllers and enhances chamber cleaning routines to eliminate contamination sources.

**Outcome:** With improved equipment reliability and cleanliness, the CVDCVD process achieves greater uniformity and reduced defect rates, enhancing overall wafer yield and product quality.

## 9.10 Equipment-related Process Defects in Wet Etch: A Hypothetical Study

**Background:** A semiconductor manufacturing facility employing wet etch processes encounters defects such as etch depth variations and pattern fidelity issues.

**Investigation:** Detailed equipment inspections reveal inconsistencies in chemical delivery systems and agitation mechanisms within the wet etch baths.

**Root cause analysis:** The defects are attributed to irregular chemical flow rates and inadequate mixing, leading to non-uniform etching.

**Corrective action:** The engineering team installs flow rate sensors and upgrades agitation mechanisms to ensure consistent chemical distribution and mixing.

**Outcome:** By optimizing equipment performance, the wet etch process achieves greater uniformity and precision, resulting in improved device performance and yield.

## 9.11 Equipment-related Process Defects in Physical Vapor Deposition (PVD): A Hypothetical Study

**Figure 9.12:** Equipment-related process defects in physical vapor deposition (PVD).

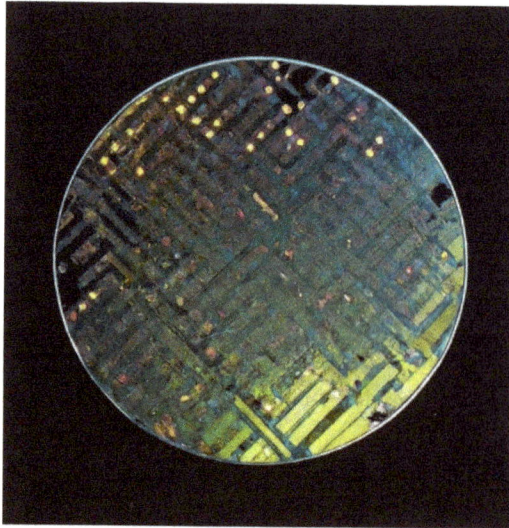

**Background:** A semiconductor fabrication facility employing PVD processes for metal deposition encounters defects such as adhesion failures and film delamination (Figure 9.12).

**Investigation:** The engineering team conducts equipment diagnostics and identifies inconsistencies in target material deposition rates and substrate temperature control.

**Root cause analysis:** The defects are traced back to uneven heating elements and target material distribution mechanisms within the PVD chambers.

**Corrective action:** The team implements heating element replacements and enhances target material distribution systems to ensure uniform deposition.

**Outcome:** With improved equipment reliability and material deposition control, the PVD process achieves enhanced adhesion and film integrity, leading to higher device reliability and yield.

## 9.12 Equipment-related Process Defects in Metrology: A Hypothetical Study

**Background:** A semiconductor manufacturing facility utilizing metrology tools for process monitoring experiences deviations in critical dimension measurements and surface roughness readings.

**Investigation:** The engineering team conducts equipment calibrations and identifies inaccuracies in sensor alignments and probe tip conditions.

**Root cause analysis:** The defects are attributed to sensor misalignments and probe tip degradation, leading to measurement inaccuracies.

**Corrective action:** The team recalibrates sensor alignments and replaces worn probe tips to ensure accurate metrology measurements.

**Outcome:** With improved equipment precision and reliability, the metrology process achieves more accurate and consistent measurements, facilitating tighter process control and improved device performance.

## 9.13 Addressing 20 Blocked Etch Defects in Semiconductor Manufacturing: A Hypothetical Case Study

**Background:** A semiconductor fabrication facility encounters recurring defects characterized by 20 blocked etch patterns, leading to yield losses and production delays.

**Equipment investigation:** In-depth analysis reveals irregularities in gas flow distribution and chamber cleanliness within the etch-equipment.

**Shift-left analysis:** By applying the shift-left approach, the engineering team identifies underlying equipment issues such as inadequate maintenance practices and improper process monitoring.

**Corrective action:** Implementing proactive maintenance protocols and real-time process monitoring systems helps mitigate gas flow inconsistencies and chamber contamination, reducing the occurrence of blocked etch defects.

**Outcome:** Through a shift-left approach to equipment management, the facility achieves improved etch process reliability and reduced defect rates, enhancing overall yield and productivity.

## 9.14 Mitigating Metal Contamination Defects in Semiconductor Manufacturing: A Hypothetical Case Study

**Background:** A semiconductor fabrication facility experiences defects related to metal contamination on semiconductor wafers, leading to device failures and yield losses.

**Equipment investigation:** Thorough examination reveals deficiencies in the deposition equipment's target material distribution and substrate handling mechanisms.

**Shift-left analysis:** Applying a shift-left approach, the engineering team identifies equipment-related issues such as inconsistent deposition rates and improper substrate handling practices (Figure 9.13).

**Figure 9.13:** An example of petty errors in semiconductor manufacturing, referred to colloquially as "minute oversights," which can significantly impact quality control and yield rates.

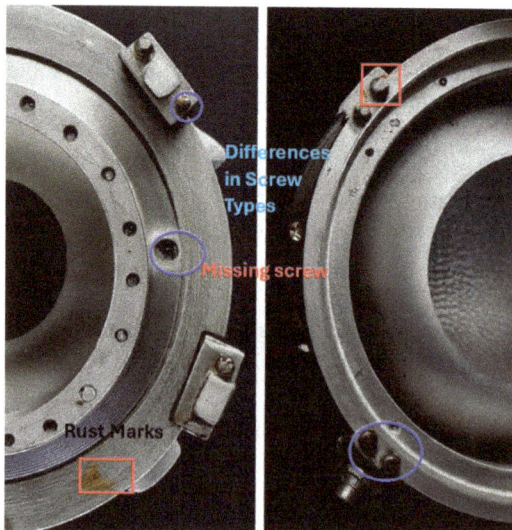

**Corrective action:** Implementing equipment upgrades and process optimizations helps ensure uniform target material distribution and precise substrate handling, minimizing the risk of metal contamination defects.

**Outcome:** By addressing equipment-related root causes through a shift-left approach, the facility achieves enhanced device reliability and yield, mitigating the impact of metal contamination defects.

## 9.15  Impact of Defects on Yield and Product Quality

Defects in semiconductor manufacturing can have significant implications for both yield and quality, impacting the overall performance and reliability of semiconductor devices. Below [46] is a detailed overview of the impact of defects on yield and quality in semiconductor manufacturing:

1.  **Yield loss**

     - **Reduced production yield:** Defects directly contribute to a decrease in production yield, as defective chips fail to meet quality standards and cannot be integrated into end products. This leads to lower overall output from the manufacturing process.
     - **Increased scrap and rework:** Defective chips must be discarded or reworked to rectify the issues, resulting in increased scrap rates and additional manufacturing costs. Rework processes also consume valuable resources and extend production cycle times.
     - **Negative financial impact:** Yield losses due to defects result in revenue loss for semiconductor manufacturers, as fewer functional chips are available for sale. This can have a significant financial impact on the company's bottom line and profitability.

2.  **Quality degradation**

     - **Device performance issues:** Defects can compromise the electrical and mechanical performance of semiconductor devices, leading to functionality issues, reduced speed, or increased power consumption. This directly affects the overall quality and reliability of electronic products [46, 71].
     - **Increased failure rates:** Devices with defects are more prone to premature failure or malfunction, resulting in higher field failure rates and potential warranty claims. This can damage the reputation of semiconductor manufacturers and erode customer trust.

- **Compatibility concerns:** Defects may cause compatibility issues with other system components or peripherals, leading to system integration challenges and customer dissatisfaction. This can result in product returns and negative reviews, impacting brand reputation.
- **Long-term reliability:** Defects can also affect the long-term reliability of semiconductor devices, leading to early wear-out or degradation over time. This can result in product recalls, costly field replacements, and liability risks for manufacturers.

3.  **Operational challenges**

- **Process disruptions:** Addressing defects requires additional time and resources, leading to production disruptions and delays. This can affect manufacturing schedules, customer delivery timelines, and overall operational efficiency.
- **Resource allocation:** Dealing with defects requires diverting resources such as manpower, equipment, and materials away from other critical tasks, impacting overall resource utilization and efficiency.
- **Quality control costs:** Implementing stringent quality control measures to detect and prevent defects incurs additional costs for semiconductor manufacturers. This includes investment in inspection equipment, testing protocols, and quality assurance processes.

In conclusion, defects in semiconductor manufacturing have far-reaching consequences on yield, quality, and operational efficiency. Addressing and mitigating defects is crucial for semiconductor manufacturers to maintain high production yield, ensure product quality and reliability, and sustain competitive advantage in the market.

## 9.16 Challenges in Traditional Defect Prevention Methods

Traditional defect prevention methods in semiconductor manufacturing face several challenges, primarily due to the complexities of the fabrication process, evolving technology nodes, and stringent quality requirements. Here's a detailed overview of the challenges associated with traditional defect prevention methods:

1.  **Limited predictive capability**

- Traditional defect prevention methods often rely on historical data and statistical analysis to identify and mitigate defects. However, these methods may lack predictive capability, making it challenging to anticipate and prevent defects in real-time.

- As semiconductor technologies advance and feature sizes shrink, defects can occur at nanoscale levels, making them difficult to detect and predict using conventional methods.

2. **Inadequate process visibility**

- Traditional defect prevention methods may provide limited visibility into the semi-conductor fabrication process, particularly at the microscopic and atomic levels where defects originate.
- Lack of comprehensive process visibility hinders the ability to identify root causes of defects and implement targeted preventive measures effectively.

3. **Complex root cause analysis**

- Identifying the root causes of defects in semiconductor manufacturing can be a complex and time-consuming process. Traditional methods such as manual inspection and statistical analysis may struggle to pinpoint the exact factors contributing to defects, especially in multi-step fabrication processes.
- Without a thorough understanding of root causes, implementing effective pre-ventive measures becomes challenging, leading to recurring defects and quality issues.

4. **High false alarm rates**

- Traditional defect prevention methods may generate a high number of false alarms or false positives, where deviations from nominal process parameters are flagged as defects erroneously.
- High false alarm rates can lead to inefficient use of resources and manpower, as operators spend time investigating and addressing non-existent defects, diverting attention from genuine issues.

5. **Difficulty in adapting to processProcess variability**

- Semiconductor fabrication processes exhibit inherent variability due to factors such as material properties, equipment performance, and environmental con-ditions. Traditional defect prevention methods may struggle to adapt to this variability, leading to inconsistent defect detection and prevention.
- Lack of flexibility in traditional methods limits their effectiveness in addressing dynamic process variations and emerging defect [35] mechanisms.

6. **Limited scalability and efficiency**

- Traditional defect prevention methods may lack scalability and efficiency, particularly in high-volume semiconductor manufacturing environments. Manual inspection and statistical analysis techniques may not be able to keep pace with the increasing complexity and throughput requirements of modern fabs.
- Scaling traditional methods to meet the demands of advanced technology nodes and large wafer sizes can pose logistical and resource challenges, hindering their effectiveness in defect prevention [36].

In summary, traditional defect prevention methods in semiconductor manufacturing face numerous challenges related to predictive capability, process visibility, root cause analysis, false alarm rates, process variability, scalability, and efficiency. Overcoming these challenges requires the adoption of advanced defect detection and prevention technologies, such as machine learning, artificial intelligence, and real-time process monitoring to enhance semiconductor yield and quality in today's highly competitive semiconductor industry.

## 9.17 Unveiling the Hidden Costs: Human Errors in Semiconductor Maintenance and their Impact on Quality and Reliability

In semiconductor manufacturing, human interaction plays a crucial role in scheduled and unscheduled maintenance activities, especially for parts requiring regular replacements and vendor interactions. Here's a detailed overview of the human factors involved and their impact on equipment reliability and quality:

1. **Scheduled maintenance and parts replacements**

- Technicians and engineers are responsible for conducting scheduled maintenance tasks, such as replacing worn-out components and performing routine inspections on equipment.
- Human error can occur during these maintenance activities, leading to issues such as improper installation of replacement parts, incorrect alignment of components, or failure to follow recommended procedures.
- Vendor interactions play a vital role in sourcing replacement parts and components. Miscommunications or errors in ordering, shipping, or receiving parts can result in delays or discrepancies, impacting equipment uptime and reliability.

2. **Unscheduled maintenance and emergency repairs**

- During unscheduled maintenance or emergency repairs, technicians must troubleshoot and address unexpected equipment failures promptly.
- Time pressure and urgency during unscheduled maintenance can increase the risk of human error, leading to incomplete repairs, misdiagnoses, or improper adjustments.
- Inadequate documentation or training on emergency procedures can further exacerbate human errors during unscheduled maintenance events.

3. **Bill of materials (BOM) complexity**

- Semiconductor equipment assemblies often involve complex BOMs with multiple components sourced from various vendors.
- Coating variations in BOM components, such as different surface treatments or material compositions, introduce additional variables that can affect equipment performance and reliability.
- Human oversight or error in selecting and specifying BOM components may result in compatibility issues, material mismatches, or suboptimal performance of equipment.

4. **Assembly and fastening errors**

- Small and seemingly insignificant issues, such as improper assembly or insufficient tightening of screws, can have significant consequences on equipment functionality and reliability.
- The use of incorrect fasteners, such as using nickel screws instead of stainless steel in corrosive environments, can lead to premature component failure or material degradation.
- Human factors such as fatigue, distraction, or lack of attention to detail can contribute to assembly errors and fastening mistakes, even in well-established manufacturing processes.

These small and petty human errors, although seemingly minor, can have catastrophic consequences on semiconductor equipment reliability and product quality. Identifying and addressing these issues require meticulous attention to detail, thorough training, robust quality control measures, and continuous improvement initiatives. Despite their elusive nature, mitigating human-induced defects is essential for ensuring the integrity and performance of semiconductor manufacturing processes.

Chapter 7 discusses a hypothetical design proposal and an innovative approach leveraging the synergies of artificial intelligence (AI), machine learning (ML), computer vision, and the Internet of Things (IoT) to transform defect detection and prevention in semiconductor manufacturing. By harnessing advanced algorithms and real-time data analytics, this design aims to identify and rectify potential errors proactively, significantly reducing the risk of costly quality defects downstream.

Within this study [20, 21, 23, 24], we delve into a theoretical framework that explores the potential of integrating artificial intelligence (AI), machine learning (ML), computer vision, and the Internet of Things (IoT) to revolutionize defect detection and prevention in semiconductor manufacturing. This conceptual approach aims to outline the theoretical benefits of proactive error detection, anticipating its potential impact on reducing downstream quality defects, while respecting the confidentiality of patent-protected designs (in progress) [1, 2, 35, 36].

**Figure 9.14:** An example of preventive maintenance on a tool.

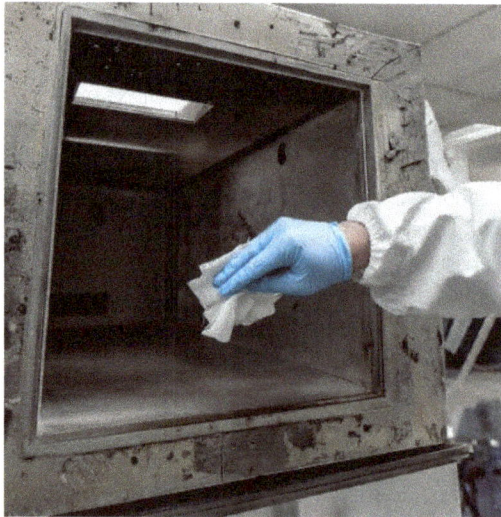

Figure 9.15: An example of preventive maintenance on a tool.

# Bibliography

[1] "SemiconductorSemiconductor ManufacturingManufacturing." Encyclopedia Britannica, Accessed February 21, 2024. https://www.britannica.com/technology/semiconductor-device.

[2] "Wafer (electronics)." Wikipedia. Wikimedia Foundation, April 11, 2023. https://en.wikipedia.org/wiki/Wafer_(electronics).

[3] Wu, Q., Li, Y., and Zhao, Y. "The Evolution of Photolithography Technology, Process Standards, and Future Outlook." In *2020 IEEE 15th International Conference on Solid-State & Integrated Circuit Technology (ICSICT)*, Kunming, China, 2020, pp. 1-5. doi: 10.1109/ICSICT49897.2020.9278164.

[4] Kanarik, K. J.; Tan, S.; Gottscho, R. A. Atomic Layer Etching: Rethinking the Art of Etch. *J. Phys. Chem. Lett.* 2018, 9(16), 4814–4821. https://doi.org/10.1021/acs.jpclett.8b00997.

[5] "Doping (semiconductor)." Simple English Wikipedia. Accessed 23 April 2023. URL: https://simple.wikipedia.org/wiki/Doping_(semiconductor).

[6] "Annealed Silicon Wafers - UniversityWafer." Accessed 23 April 2023. https://www.universitywafer.com/annealed-silicon-wafer.html.

[7] City University of Hong Kong. "5. Polarization." Accessed from https://www.cityu.edu.hk/phy/appkchu/AP6120/5.PDF.

[8] Rossnagel, S. "Thin film deposition with physical vapor deposition and related technologies." *Journal of Vacuum Science & Technology A - J VAC SCI TECHNOL A* 21 (September 1, 2003). doi: 10.1116/1.1600450.

[9] M. A. Lieberman and A. J. Lichtenberg, Principles of Plasma Discharges and Materials Processing, 2nd ed. (John Wiley & Sons,Inc., New York, 2005).

[10] AnandTech. "ASML to Ship Multiple High-NA Tools in 2025, Expands Production Capacities." AnandTech, accessed February 22, 2024. https://www.anandtech.com/show/21264/asml-to-ship-multiple-highna-tools-in-2025-expands-production-capacities.

[11]    "Semiconductor Front-End Process Episode 6 - SK hynix Newsroom." Accessed 23 April 2023. https://news.skhynix.com/semiconductor-front-end-proce ss-episode-6/.

[12]    "Review of the advances in low-cost silicon production technologies with reduced carbon-emission." Scientific Figure. ResearchGate. Accessed April 24, 2023. https://www.researchgate.net/figure/A-timeline-of-the-early-developm ent-of-silicon-extraction-by-electrochemistry_fig2_370197747.

[13]    Demir, H.V., Jun-Fei Zheng, V.A. Sabnis, Onur Fidaner, J. Hanberg, James Harris, and David Miller. "Self-Aligning Planarization and Passivation for Integration Applications in III–V Semiconductor Devices." Semiconductor Manufacturing, IEEE Transactions on 18, no. 2 (March 1, 2005): 182–189. https://doi.org/10.1109/TSM.2004.841834.

[14]    Ibrahim, A. M. Y., and H. G. Kerkhoff. "DARS: An EDA Framework for Reliability and Functional Safety Management of System-on-Chips." In 2019 IEEE International Test Conference (ITC), pp. 1-10. Washington, DC, USA: IEEE, 2019. https://doi.org/10.1109/ITC44170.2019.9000112.

[15]    Lei, Wei-Sheng, Ajay Kumar, and Rao Yalamanchili. "Die singulation technologies for advanced packaging: A critical review." J. Vac. Sci. Technol. B 30 (2012): 040801. https://doi.org/10.1116/1.3700230.

[16]    Oricus Semiconductor. "What is the Die Attach Process?" Accessed April 24, 2024. https://oricus-semicon.com/what-is-the-die-attach-process/.

[17]    Khoury, S.L., Burkhard, D.J., Galloway, D.P., & Scharr, T.A. "A comparison of copper and gold wire bonding on integrated circuit devices." In Proceedings of the Electronic Components and Technology Conference (ECTC), 768–776. 1990. doi: 10.1109/ECTC.1990.122277. S2CID 111130335. Retrieved February 6, 2024.

[18]    SensXpert. "Essentials of Electronics Encapsulation." SensXpert Blog. Accessed on April 23, 2024. https://www.sensxpert.com/blog/essentials-electronics-en capsulation/.

[19]    Accel-RF Instruments Corporation. "Semiconductor Reliability Testing Guide." Accessed April 24, 2024. https://info.accelrf.com/semiconductor-reliability-t esting-guide.

[20]    Ucar, Aysegul, Mehmet Karakose, and Necim Kırımça. 2024. "Artificial Intelligence for Predictive Maintenance Applications: Key Components, Trustworthiness, and Future Trends" Applied Sciences 14, no. 2: 898, https: //doi.org/10.3390/app14020898.

[21]    Pheng, M. S. K., and David, L. G. "Artificial Intelligence in Back-End Semiconductor Manufacturing: A Case Study." In 2022 IEEE International Conference on Distributed Computing and Electrical Circuits and Electronics (ICDCECE), Ballari, India, 2022, pp. 1-4. doi: 10.1109/ICDCECE5 3908.2022.9792976.

[22]  Y.-R. Yeh, Z.-Y. Lee, and Y.-J. Lee, "Anomaly detection via over-sampling principal component analysis," in New Advances in Intelligent Decision Technologies. Cham, Switzerland: Springer, 2009, pp. 449–458.

[23]  C. -F. Chien, H. Ehm, J. W. Fowler, K. G. Kempf, L. Mönch and C. -H. Wu, "Production-Level Artificial Intelligence Applications in Semicon-ductorSemiconductor Supply Chains," in *IEEE Transactions on Semiconductor Manufacturing*, vol. 36, no. 4, pp. 560-569, Nov. 2023, doi: 10.1109/TSM.2023.3322142.

[24]  Lin, Y.-C.; Yeh, C.-C.; Chen, W.-H.; Hsu, K.-Y. Implementation Criteria for Intelligent Systems in Motor Production Line Process Management. *Processes* 2020, 8, 537.

[25]  Hurtarte, J.S., Wolsheimer, E.A., and Tafoya, L.M. "Semiconductor Manufacturing Basics." In *Understanding Fabless IC Technology*, Chapter 4, pp. 41–45. Burlington, MA, USA: Newnes, 2007. ISBN 978-0-7506-7944-2.

[26]  Espadinha-Cruz, Pedro, Radu Godina, and Eduardo M. G. Rodrigues. 2021. "A Review of Data Mining Applications in SemiconductorSemiconductor Manufacturing" *Processes* 9, no. 2: 305. https://doi.org/10.3390/pr9020305.

[27]  Lee, Youjin, and Yonghan Roh. "An Expandable Yield Prediction Framework Using Explainable Artificial Intelligence for Semiconductor Manufacturing." *Applied Sciences* 13, no. 4 (2023): 2660. https://doi.org/10.3390/app13042660.

[28]  J. Jiang, W. Lin, and N. Raghavan, "A Novel Framework for Semiconductor Manufacturing Final Test Yield Classification Using Machine Learning Techniques," *IEEE Reliability Society Section*, vol. 1, no. 1, pp. 1-10, Oct. 29, 2020, doi: 10.1109/ACCESS.2020.3034680.

[29]  G. N. Silveira, R. F. Viana, M. J. Lima, H. C. Kuhn, C. D. P. Crovato, S. B. Ferreira, G. Pesenti, E. Storck, and R. R. Righi, "I4.0 Pilot Project on a Semiconductor Industry: Implementation and Lessons Learned," *Sensors*, vol. 20, no. 20, p. 5752, Oct. 10, 2020, doi: 10.3390/s20205752.

[30]  D. Kim and K. Cho, "Digital Transformation Characteristics of the Semiconductor Industry Ecosystem," *Sustainability*, vol. 15, no. 1, p. 483, 2023, doi: 10.3390/su15010483.

[31]  "Chapter 3: Internet of Things (IoT)." *IEEE Electronics Packaging Society*, 2019, http://eps.ieee.org/hir/chapter3.

[32]  Maciej Kranz, Building the Internet of Things: Implement New Business Models, Disrupt Competitors, Transform Your Industry (Wiley, 2016).

[33]  Ojuawo, Olutayo, and Folahan Jiboku. "Overview of Edge Computing and Its Significance in the Era of IoT and Big Data." September 20, 2023.

[34]  Furht, B., & Villanustre, F. (2016). *Introduction to Big Data. In: Big Data Technologies and Applications*. https://doi.org/10.1007/978-3-319-44550-2_1.

[35]  Alkauskas, Audrius, Matthew D. McCluskey, and Chris G. Van de Walle. "Tutorial: DefectsDefects in Semiconductors—Combining Experiment and Theory." Journal of Applied Physics 119, no. 18 (2016): 181101. doi: 10.1063/1.4948245. Accessed from https://doi.org/10.1063/1.4948245.

[36]  Huang, Y.-H., & Tsai, C.-M. (2014). Sources of defects in semiconductor manufacturing. IEEE Transactions on Semiconductor Manufacturing, 27(3), 403-415. https://doi.org/10.1109/TSM.2014.2329691.

[37]  Bathen, M. E., C. T.-K. Lew, J. Woerle, C. Dorfer, U. Grossner, S. Castelletto, and B. C. Johnson. "Characterization Methods for DefectsDefects and Devices in SiliconSilicon Carbide." Journal of Applied Physics 131, no. 14 (2022): 140903. doi: 10.1063/5.0077299.

[38]  Drabold, David A., and Stefan K. Estreicher, eds. Theory of Defects in Semiconductors. Springer.

[39]  C. Bergès, J. Bird, M. D. Shroff, R. Rongen and C. Smith, "Data analytics and machine learning: root-cause problem-solving approach to prevent yieldyield loss and quality issues in semiconductor industry for automotive applications," 2021 IEEE International Symposium on the Physical and Failure Analysis of Integrated Circuits (IPFA), Singapore, Singapore, 2021, pp. 1-10, doi: 10.1109/IPFA53173.2021.9617238. https://doi.org/10.1109/TSM.2018.2838631.

[40]  M. Komagata, "A new method of reducing the particle contamination in semiconductor manufacturing," Proceedings of 1995 Japan International Electronic Manufacturing Technology Symposium, Omiya, Japan, 1995, pp. 146-149, doi: 10.1109/IEMT.1995.541014.

[41]  Kommisetti, Subrahmanyam, Scott Singlevich, Paul Ewing, and Michael Johnson. "Detecting Arcing Events in Semiconductor Manufacturing Equipment." IEEE Transactions on Semiconductor Manufacturing 26 (November 1, 2013): 488-492. doi: 10.1109/TSM.2013.2283053.

[42]  O.Auciello and D. L.Flamm, Plasma Diagnostics: Discharge Parameters and Chemistry (Academic, New York, 1989).

[43]  LithoGuru. "Papers." Accessed from https://www.lithoguru.com/scientist/papers.html.

[44]  Li, Qian-Kun, Xiao, Yao, Liu, Hua, Zhang, Hao-Lin, Xu, Jia, and Li, Jin-Huan. "Analysis and correction of the distortion error in a DMD based scanning lithography system." Optics Communications 434 (2019): 1-6. ISSN 0030-4018. doi: 10.1016/j.optcom.2018.10.042.

[45]  Ching-Fa Yeh et al., "The removal of airborne molecular contamination in cleanroom using PTFE and chemicalchemical filters," in IEEE Transactions on Semiconductor Manufacturing, vol. 17, no. 2, pp. 214-220, May 2004, doi: 10.1109/TSM.2004.826957.

[46] Semiconductor Industry Association. "Yield Management in Semiconductor Manufacturing." Accessed on April 23, 2024. https://www.semiconductors.org/wp-content/uploads/2018/09/Yield.pdf.

[47] AnandTech. "Intel Articles." AnandTech, accessed February 22, 2024. https://www.anandtech.com/tag/intel.

[48] "Advanced Solution for Critical Layers and Supports 3D Semiconductor Device Production." December 6, 2023.

[49] KLA Corporation. "Defect Inspection & Review | KLA." Accessed February 21, 2024. https://www.kla.com/products/chip-manufacturing/defectdefect-inspection-review.

[50] Lam Research Corporation. "EtchEtch | Lam Research." Accessed February 21, 2024. https://www.lamresearch.com/products/our-processes/etch/.

[51] Applied Materials, Inc. "Deposition Technologies for Every Device." Accessed February 21, 2024. https://www.appliedmaterials.com/us/en/semiconductor/products/processes/ald.html.

[52] NVIDIA Corporation. "NVIDIA, ASML, TSMC, and Synopsys Set Foundation for Next-Generation Chip Manufacturing." Accessed February 21, 2024. https://nvidianews.nvidia.com/news/nvidia-asml-tsmc-and-synopsys-set-foundation-for-next-generation-chip-manufacturing.

[53] Techcet. "GLOBALFOUNDRIES Rolls Out 12nm FD-SOI Process." Accessed February 21, 2024. https://techcet.com/globalfoundries-rolls-out-12nm-fd-soi-process/.

[54] GlobalFoundries. "Leading Semiconductor Players to Advance Next-Generation FD-SOI Roadmap for Automotive, IoT, and Mobile Applications." Accessed February 21, 2024. https://gf.com/gf-press-release/leading-semiconductor-players-to-advance-next-generation-fd-soi-roadmap-for-automotive-iot-and-mobile-applications/.

[55] GLOBALFOUNDRIES. "GLOBALFOUNDRIES Surpasses $2 Billion in Design Win Revenue on 22FDX® Technology." Press release, July 09, 2018.

[56] "Samsung Begins Chip Production Using 3nm Process Technology With GAA Architecture." Korea, June 30, 2022. Accessed February 21, 2024. https://news.samsung.com/global/samsung-begins-chip-production-using-3nm-process-technology-with-gaa-architecture.

[57] https://doi.org/10.1016/S0042-207X(99)00189-X - P.J Kelly, R.D Arnell, Magnetron sputtering: a review of recent developments and applications, Vacuum, Volume 56, Issue 3, 2000, Pages 159-172, ISSN 0042-207X.

[58] https://doi.org/10.1116/1.4998940 - E. Greene, J. (2017). Review Article: Tracing the recorded history of thin-film sputter deposition: From the 1800s to 2017. Journal of Vacuum Science & Technology A: Vacuum, Surfaces, and Films. 35.

[59]   https://doi.org/10.1016/B978-0-8155-2031-3.00005-3 - D. Depla, S. Mahieu, J.E. Greene, Chapter 5 - Sputter Deposition Processes, Editor(s): Peter M. Martin, Handbook of Deposition Technologies for Films and Coatings (Third Edition), William Andrew Publishing, 2010, Pages 253-296.

[60]   Nahir, Amir, Avi Ziv, R. Galivanche, Alan Hu, Miron Abramovici, Albert Camilleri, Bob Bentley, Harry Foster, Valeria Bertacco, and Shakti Kapoor. "Bridging Pre-Silicon Verification and Post-Silicon Validation." 94-95. 2010. doi: 10.1145/1837274.1837300.

[61]   "Vivado." Wikipedia. Wikimedia Foundation, February 20, 2024. https://en.wikipedia.org/wiki/Vivado.

[62]   ASE Technology Holding Co., Ltd. "ASE Global." Accessed on March 25, 2024. https://www.aseglobal.com/.

[63]   Wikipedia contributors. "Qualcomm." Wikipedia, The Free Encyclopedia. Wikimedia Foundation, Inc. January 25, 2022. Web. March 25, 2024. https://en.wikipedia.org/wiki/Qualcomm.

[64]   Wikipedia contributors. "Amkor Technology." Wikipedia, The Free Encyclopedia. Wikimedia Foundation, Inc. March 9, 2022. Web. March 25, 2024. https://en.wikipedia.org/wiki/Amkor_Technology.

[65]   Wikipedia contributors. "Texas Instruments." Wikipedia, The Free Encyclopedia. Wikimedia Foundation, Inc. March 19, 2024. Web. March 25, 2024. https://en.wikipedia.org/wiki/Texas_Instruments.

[66]   Williams, T. W., and Parker, K. P. "Design for Testability—A Survey." Proceedings of the IEEE 71, no. 1 (Jan. 1983): 98-112. doi: 10.1109/PROC.1983.12531.

[67]   IEEE Spectrum. "Chip Design." Accessed April 24, 2024. https://spectrum.ieee.org/tag/chip-design.

[68]   "Electronic Design Automation." Wikipedia. Accessed April 24, 2024. https://en.wikipedia.org/wiki/Electronic_design_automation.

[69]   "Pentium FDIV bug." Wikipedia. Accessed April 24, 2024. https://en.wikipedia.org/wiki/Pentium_FDIV_bug.

[70]   Heterogeneous Integration Roadmap 2023, ed. Chapter 17, Page ii. Accessed from eps.ieee.org/hir.

[71]   Stich, P., Wahl, M., Czerner, P., Weber, C., and Fathi, M. "Yield prediction in semiconductor manufacturing using an AIAI-based cascading classification system." In 2020 IEEE International Conference on Electro Information Technology (EIT), Chicago, IL, USA, 2020, pp. 609-614. doi: 10.1109/EIT48999.2020.9208250.

[72]   Chen, Erli. "Fabrication II - Deposition." Accessed from https://www.mrsec.harvard.edu/education/ap298r2004/Erli%20chenFabrication%20II%20-%20Deposition-1.pdf.

[73]   "Stoney Equation." ScienceDirect. Accessed from https://www.sciencedirect.com/topics/engineering/stoney-equation#:~:text=Stoney's%20equation%20is%20used%20to,fr

om%20different%20molecular%20recognition%20events.&text=where%20R%20is%20the%2 0radius,is%20the%20surface%20stress%20generated.

[74]    "MicroelectromechanicalMicroelectromechanical Systems." Wikipedia. Accessed from https://en.wikipedia.org/wiki/MEMSMEMS.

[75]    Antunes, J.M., Fernandes, J.V., Sakharova, N.A., Oliveira, M.C., and Menezes, L.F. "On the determination of the Young's modulus of thin films using indentation tests." *International Journal of Solids and Structures* 44, no. 25–26 (2007): 8313-8334. ISSN 0020-7683. doi: 10.1016/j.ijsolstr.2007.06.015.

[76]    Katskov, Dmitri, and Nicholas Darangwa. "Application of Langmuir theory of evaporation to the simulation of sample vapor composition and release rate in graphite tube atomizers. Part 1. The model and calculation algorithm." *Journal of Analytical Atomic Spectrometry - J ANAL ATOM SPECTROM* 25, no. 1 (June 22, 2010). doi: 10.1039/c002017f.

[77]    "Chemical VaporVapor Deposition." ScienceDirect. Accessed from https:// www.sciencedirect.com/topics/materials-science/chemicalchemical-vapor-depositiondeposition.

[78]    T. C.Penn, IEEE Trans. Electron. Dev.ED26, 640 (1979). https: //doi.org/10.1109/T-ED.1979.19471.

[79]    University of Münster. "Heat Conduction Equation." Accessed 23 April 2023. https://www.uni-muenster.de/imperia/md/content/physik_tp/lectures/ws2016-2017/num _methods_i/heatheat.pdf.

[80]    "Fick's laws of diffusion." Wikipedia. Wikimedia Foundation, https://en.wikipedia.org/wiki/Fick%27s_laws_of_diffusion.

[81]    Tektronix. "ABCs of Probes Primer." Accessed from chrome-extension://efaidn bmnnnibpcajpcglclefindmkaj/https://web.mit.edu/6.101/www/reference/ABCprobes_s.pdf.

# Index

fabrication 4, 18, 31, 51, 63, 70, 101, 103, 116
fabs 18, 20, 122
facilities 18, 20, 99, 100
feature 24, 40, 52, 54, 65, 109
flow 8, 51, 63, 106, 108, 117
foundries 18, 19, 20
foundry 11, 21, 27
functional testing 14

**G**
gas constant 45, 50, 57
gases 32, 43, 52, 53

**H**
heat 39, 64, 74, 106
human 28, 73, 82, 90, 122

**I**
idm 18, 19, 20, 21
illumination 66
in-circuit test 15
inconsistencies 91, 106, 111, 115, 117
ingot 6, 7
integrated device manufacturer 19, 20
interface 77
intrinsic 59, 102
ionized 40
iot 1, 27, 91, 92, 97, 100
isotropic 51, 53, 54, 67, 78

**J**
journey 2, 5, 8

**L**
laser 75, 103
layout 11, 13, 47
lithography 18, 24, 27, 66, 77, 109

**M**
magnetron sputtering 32, 41, 42, 131
manufacturing 4, 20, 23, 31, 65, 77

mask 24, 48, 57, 67
masks 51, 66, 109, 112
material 2, 24, 31, 34, 50, 69, 71, 118
mems 36, 37, 52
metal 31, 44, 64, 70, 78, 105, 110
metallization 18, 64, 78
metrology 24, 67, 103, 117
micro-arcing 105, 106, 108
microelectromechanical 36, 52
molar mass 35, 37
moore's law 16, 24

**N**
notation 101
numerical aperture 66

**O**
osat 19, 21
outsourced 19, 21
oxide 4, 63, 70, 111

**P**
params 67, 71
particle 104, 105, 130
photoresist 24, 46, 47, 49, 66, 102
pitch 24
plasma 26, 32, 33, 44, 52, 54
polarities 42
post-silicon 1, 8, 14, 15, 16
precursor 32, 33, 35, 43, 45
predictive maintenance 81, 83, 85, 89, 94, 100
pre-silicon 1, 8, 9, 14, 15, 16
pressure 34, 38, 55, 86
preventive maintenance 81, 83, 84, 85, 110, 114, 124, 125
probing 73, 74, 75
process 1, 8, 23, 31, 42, 52, 65, 80, 104, 116
profile 51, 60, 67
protocols 82, 83, 110

For Product Safety Concerns and Information please contact our EU
representative  GPSR@taylorandfrancis.com
Taylor & Francis Verlag GmbH, Kaufingerstraße 24, 80331 München, Germany

www.ingramcontent.com/pod-product-compliance
Lightning Source LLC
Chambersburg PA
CBHW070732220326
41598CB00024BA/3400